SpringerBriefs in Physics

D1550460

More information about this series at http://www.springer.com/series/8902

Alexander Herega

The Selected Models of the Mesostructure of Composites

Percolation, Clusters, and Force Fields

 Springer

Alexander Herega
Department of Computer Systems
Odessa National Academy of Food
  Technologies
Odessa
Ukraine

ISSN 2191-5423   ISSN 2191-5431   (electronic)
SpringerBriefs in Physics
ISBN 978-3-319-89703-5   ISBN 978-3-319-89704-2   (eBook)
https://doi.org/10.1007/978-3-319-89704-2

Library of Congress Control Number: 2018940889

Printed on acid-free paper

This Springer imprint is published by the registered company Springer International Publishing AG
part of Springer Nature
The registered company address is: Gewerbestrasse 11, 6330 Cham, Switzerland

Preface

It is generally accepted that the most habitual impression of the surrounding world is a huge number of objects and a huge variety of materials from which they are made. Polyfunctional composite materials which were developed in recent decades meet manifold requirements. What is more, they have peculiar, sometimes unique properties. Creation of such materials is based on the understanding of laws of the interrelation between structure and properties, and on the structural theory of material.

There were different stages in the establishment of Material Science as a fundamental science: the time of the emergence of new theories and impressive breakthroughs in the understanding, periods of the scarcity of ideas and concepts, the aggregation of unresolved problems. There were changes and interweaving of stages of terminology formation and the creation of experimental and calculated methods. Also periods of prevailing paradigms were clearly observed during the development of Material Science.

The modern stage of the development of Physical Material Science differs by certain balance: There is no dominating concept now and researchers' attention is divided between the theory of continuous media, the system of views of the theory of condensed state and the research, the subject of which is phenomena caused by mesostructure of material.

As any developing system, solids have a hierarchical structure—a set of different scale levels of the organization of the structure that interact. Mesomechanics is the scientific direction of solid-state physics, which views its structure as a multilevel self-organizing system at the micro-, meso-, and macroscopic scales. Mesoscopic scale (or mesoscopic asymptotics, or intermediate asymptotic) can be defined as an interval of dimensions in which structural elements are statistically self-similar. Two systems could be called as statistically self-similar if they have equal mathematical expectations and the same type of statistical distribution. The characteristic values of the intermediate asymptotic dimensions cannot be indicated, since, depending on the type of a system they can be in any area of scale—from nanometers to centimeters. Also for the intermediate asymptotic is typical, in particular, the presence of the scale invariance in a certain range of dimensions, as well as the inability to averaging of bodies properties by realizations.

Modification of the mesostructure is one of the rapidly developing areas of the solids' genesis. A purposeful effect on the mesostructure, which can be realized through the influence on processes of self-organization, requires the understanding of features of the origin and evolution of homogeneous areas in the material, knowledge of the nature of their interaction and influence on the properties of composite materials, the study of the cluster system of the material, and the study of properties of phase transitions occurring at the critical concentration of clusters, etc.

One of the research areas of Material Science is composites. Physical encyclopedia defines them as heterogeneous thermodynamically disequilibrium systems consisting of two or more components, each of which is incorporated into a material in order to give it required properties; at the same, ingredients are different in chemical composition, physical, mechanical, and other properties, and as well, they are separated by clearly defined boundary.

It is known that the thermodynamic disequilibrium of composite materials is connected with the ability to change their state without energy exchange with the environment as a result of the interaction between parts of a material. The presence of sharp boundaries is a consequence of the action of factors that limit a degree of intensity of the interaction between components. Undoubtedly, the interaction is inevitable and necessary. However, the intensive interaction leads to the mutual dissolution of components and to the emergence of intermediate phases.

In the researches that compile the book, various aspects of the influence of the structure on properties of the material (e.g., conductivity, crack resistance, elasticity) are considered, some patterns that occur in the initial stage of composites' formation, especially processes of the genesis and the evolution of interior boundaries of the material, issues of cooperation between cluster structures as well as other issues are described.

It is a pleasure to thank colleagues for the cooperation and support, as well as for constructive discussions on many issues considered in the monograph. Thank you very much to Dr. Olga Titlova for your help in translation into English.

Odessa, Ukraine   Alexander Herega

Contents

Abstract

Problems of interrelation between structure and properties of fractal formations of composite materials are considered in the monograph. Formation processes and properties of fractal aggregates, interior boundaries of the material, percolation clusters of phases, and other elements' structures are studied analytically and by computer simulations. A technique for a quantitative description of inhomogeneous mediums is proposed. The book is intended for researchers and HEIs professors, postgraduate, and senior students of physical, technical, and engineering faculties.

Chapter 1
Introduction

Regular research of materials is conducted in Europe, approximately, from the middle of the seventeenth century. Formed by then conceptions of the nature of matter and methods of its study, the high level of Mathematics' development and vigorous establishment of Natural Science have led to domination during two centuries of the continuum model of matter [1–8]. Phenomenological theories based on this model still find many applications in various areas of Mechanics, Physics, Technical and Material Science. According to this theory, matter can be regarded as a continuous medium, neglecting its molecular structure and can be described with convention averages: density, strength, mechanical stress and others playing the role of field variables, which allows using the apparatus of mathematical analysis, in particular integral and differential calculus.

The continual approach made a great contribution to the description of nature. It is a basis for numerous scientific and engineering theories, models and methods of calculations and is still one of the key in the study of matter. It was only, approximately, the beginning of the nineteenth century when the concept, according to which properties of materials are determined by their structure, was approved in scientific Materials Science.

In modern Natural Science the notion of structure is one of the basic. It can be understood as «a model description that sets out the order in the distribution of characteristic quantities» [9] or as a «condition that occurs as a result of the coherent behavior of a large number of particles» [10].

Definitions of this type do not claim to be complete and cannot be exhaustive: the structure as a fundamental notion is defined by a list of properties that contains both quantitative (lattice constants, coordination numbers, binding energy, etc.) and qualitative (a kind of lattice, a type of order, an availability of symmetry, etc.) characteristics. One of the structure components—clusters.

For the first time they were mentioned in the book «The Sceptical Chymist» written by Robert Boyle and published in 1661 as «groups that cannot easily be divided into their constituent parts» [11]. In modern Chemistry the notion of the cluster has appeared only in the 60s of the 20th century: a union of a finite number of metal atoms held down in a great extent by interior connections [12]. Interest to such structures has led to the emergence of a new scientific direction—chemistry of cluster compounds. Widespread specific objects with «profound structural features» [13],

© The Author(s), under exclusive licence to Springer International Publishing, part of Springer Nature 2018
A. Herega, *The Selected Models of the Mesostructure of Composites*, SpringerBriefs in Physics, https://doi.org/10.1007/978-3-319-89704-2_1

peculiar reactionary abilities and characteristic areas of application have become the subjects of study for this direction. A profound analogy between the cluster compounds of metal and cage compounds of non-metals has been elucidated. This allowed concluding that molecular compounds with backbones in the form of chains, cycles of carcasses or polyhedrons formed by atoms of a given element [14] can be obtained for every chemical element. Additionally, it made possible to understand that the Cluster Compounds Chemistry can be considered as «Organic Chemistry» of non-carbon elements [13]. In other words, from the examination of the structure, electronic structure, energetics of known cluster compounds does not imply any fundamental prohibitions of the fact that on the basis of any element of the periodic system molecular compounds with all the elementary structural elements typical for carbon compounds were realized [14].

It is generally accepted that materials scientists and physicists consider cluster as the system of a finite number of bound atoms or molecules, or as the group of small particles of finite size that have certain properties: fractal structure, considerable specific surface area, extensive density and magic numbers for the quantity of particles [15–18]. This distinguishes clusters from the «regular» set of particles forming a kind of group with equally remarkable but other properties [19, 20].

Fractal clusters are main structure-forming elements of many macroscopic systems, arising in variety of physical and chemical processes [21]. For instance, during the formation of gel in a solution, the relaxation of a metal steam in fumes and mists, the formation of pellicles on a surface during their spraying from the jet containing aerosols [14] and others.

It is interesting that the availability of cluster aggregates in the structure can lead to a new physical state of matter. For example, aerogels with their properties are intermediate between gas and condensed body and they are independent physical state of matter [22]. Such structures have a particular set of physical-chemical properties [23]: very low density, developed pore structure and high specific surface area which makes them attractive for solutions of a number of applied problems.

Experimental research of clusters is hampered by complexity and laboriousness of experiments. The key problem is that in order to obtain reliable information on the structure and physical-chemical properties it is necessary to synthesize aggregates with a wide range of parameters and to create a reliable system for their diagnostics.

Due to the difficulties experienced by experimenters at clusters' study [9], great hopes are placed on theoretical studies, including imitational and statistical simulation. Furthermore, «apparently, the most reliable prediction of the properties of such systems is provided only by machine calculations conducted, in particular, by the Monte Carlo method» [19] (see Appendix 1).

Section of the probability theory, which has numerous applications in natural and engineering sciences,—the percolation theory—has been studying for over half a century the features of the emergence and the evolution, as well as properties of clusters in matrixes of the arbitrary structure and dimension [24–26]. The theory is used in Physics of solids to describe the effects in inhomogeneous mediums in the event of connected (or quasiconnected) areas which have characteristic sizes comparable to dimensions of the body—percolation clusters of defects, the phases of matter, separation boundaries and others.

A sustained interest of specialists in percolation clusters observed over recent decades is explained by the obvious importance of the study of critical phenomena: near the phase transition point, due to the large size of clusters, the geometry of the system does not depend on the type of material in many ways and has universal properties inherent in systems of different nature [24–26]. Percolation clusters, which are formed, cardinally modify the material: as a result of the structural phase transition the correlation function undergoes a discontinuity of the second kind, the symmetry of the object is reduced, and other parameters are changed too. All these lead to the transformation of physical-chemical and mechanical characteristics of physical objects. These clusters significantly modify conductivity processes, influence on the kinetics of chemical reactions, determine the mechanical strength and the corrosion resistance, lead to anomalous diffusion and other phenomena. Therefore, in percolation studies the cluster system of the physical body and its influence on the object as a whole are usually studied simultaneously [24–28].

References

1. Timoshenko, S.P.: History of Strength of Materials: With a Brief Account of the History of Theory of Elasticity and Theory of Structures. McGraw-Hill Book Company, New York (1953)
2. Sedov, L.I.: Similarity and Dimensional Methods in Mechanics. CRC Press, Boca Raton, Florida (1993)
3. Landau, L.D., Lifshitz E.M.: Theory of Elasticity. Pergamon Press, New York (1989)
4. Ovsyannikov, L.V.: Vvedeniye v mekhaniku sploshnykh sred. Chasti 1, 2. (An Introduction to Continuum Mechanics. Parts 1, 2). Novosibirsk University Press, Novosibirsk (1976–77)
5. Ashcroft, N.W., Mermin, N.D.: Solid State Physics. Holt, Rinehart and Winston, New York (1976)
6. Wert, C.A., Thomson, R.M.: Physics of Solids. McGraw-Hill, New York (1964)
7. de Saint-Venant, B.: Memuar o izgibe krucheniya prizm. (Memoir on Prism Torsion Bending). Fizmatlit, Moscow (1961)
8. Zhukovsky, N.E.: Teoreticheskaya mekhanika. (Theoretical mechanics). GITTL, Moscow (1952)
9. Samarskiy, A.A., Zmitrenko, N.V., Kurdyumov, S.P., Mikhailov, A.P.: Thermal structures and fundamental length in a medium with nonlinear heat conduction and volumetric heat sources. Sov. Phys. Dokl. 21, 141–143 (1976)
10. Haken, H.: Information and Self-Organization: A Macroscopic Approach to Complex Systems. Springer, New York (2006)
11. Lakhno, V.D.: Klastery v fizike, khimii, biologii. (Clusters in physics, chemistry, biology). R&H Dynamics, Moscow- Izhevsk (2001)
12. Wilkinson, G., Cotton, F.A.: Progress in Inorganic Chemistry 1, 1–124 (1959)
13. Gubin, S.P.: Khimiya klasterov. Osnovy klassifikatsii i stroenie (Chemistry of Clusters. Fundamentals of Classification and Structures). Nauka, Moscow (1987)
14. Gubin, S.P.: Khimiya klasternykh soyedineniy—novoye nauchnoye napravleniye (Chemistry of cluster compounds—a new scientific direction)..Bulletin of the Academy of Sciences of the USSR 1, 55–59 (1984)
15. Mandelbrot, B.: The Fractal Geometry of Nature. W.H. Freeman and Co., San Francisco (1982)

16. Smirnov, B.M.: Properties of a fractal aggregate. Sov. Phys. Usp. **32**, 181–182 (1989). https://doi.org/10.1070/PU1989v032n02ABEH002682
17. Smirnov, B.M.: Fractal clusters. Sov. Phys. Usp. **29**, 481–505 (1986). https://doi.org/10.1070/PU1986v029n06ABEH003414
18. Feder, J.: Fractals. Plenum Press, New York (1988)
19. Petrov, Y.N.: Fizika malykh chastits (Physics of Small Particles). Nauka, Moscow (1982)
20. Stace, T.: How small is a solid? Nature **331**, 116–117 (1988)
21. Pietronero, L., Tosatti E. (ed.): Fractals in Physics. North-Holland, Amsterdam, New York (1986)
22. Lushnikov, A.A., Negin, A.E., Pakhomov, A.V., Smirnov, B.M.: Aerogel structures in a gas. Sov. Phys. Usp. **34**(2), 160–170 (1991)
23. Smirnov, B.M.: Aerogels. Sov. Phys. Usp. **30**, 420–432 (1987). https://doi.org/10.1070/PU1987v030n05ABEH002906
24. Sokolov, I.M.: Dimensions and other critical indictors in the percolation theory. Sov. Phys. Usp. **29**, 924-945 (1986). https://doi.org/10.1070/PU1986v029n10ABEH003526
25. Shklovskii, B.I., Efros, A.L.: Electronic Properties of Doped Semiconductors. Springer, Berlin (1984)
26. Efros, A.L.: Physics and Geometry of Disorder. Percolation Theory. MIR Publishers, Moscow (1986)
27. Herega, A., Sukhanov, V., Vyrovoy, V.: The Model of the Long-Range Effect in Solids: Evolution of Structure Clusters of Interior Boundaries, and Their Statistical Descriptors. AIP Conference Proceedings 1909, 020069 (2017)
28. Herega, A.N.: Physical aspects of self-organization processes in composites. 1. Simulation of percolation clusters of phases and of inner boundaries. Nanomechanics Sci. Technol. **4**(2), 119–132 (2013)

Chapter 2
Complex Percolation Model of the Material Structure

Abstract To solve the problems of physical material science, in particular, to study the structure of the material at the mesoscopic scale, to study the structure and properties of percolation clusters in the volume and on the surface of solids, the multi-functional percolation model has been developed. It is designed to study percolation clusters constructed by the Monte Carlo method using a wide range of algorithms, which allows using it for studying various phenomena and processes, including for studying of composites whose structural properties are predominantly formed with the help of connected regions. Depending on the purpose of the study, the role of percolation clusters in the model is played by various structural elements: internal boundaries, the accumulation of a different nature defects, a group of particles, phases of matter, cracks, and others. The basis of its computer implementation is an object-oriented approach, which makes it convenient to vary the objects of research, change their properties and conditions for carrying out computer experiments. The monograph introduces the idea of the Sierpinski carpet with a hybrid (finite–infinite) branching; the continuum percolation problem is considered, and the characteristics of the percolation transition on such a carpet are calculated.

In the percolation theory development, the monograph suggests the concept of the immediate neighborhood of cluster elements. In this approach, the immediate neighborhood is a percolation field of smaller scale, for which, as in case of percolation problems in general, algorithms for determining the measure (in a sense of the dimension theory) in the set of conducting sections and the dimension describing the scaling behavior of the entropy of their decomposition are represented; an idea of a relative degree of ordering of the structure is introduced, and a suitability of this quantity for estimating a drift of the immediate neighborhood properties is shown.

The "percolation boom" of the 70s of the last century showed that percolation theory has so many modifications of algorithms for constructing connected domains that there is a "legion of them" [1–8].

A multifunctional percolation model to study the material structure in intermediate asymptotics, what is more, the structure and properties of percolation clusters in the volume and on the surface of solids was developed in [8–14]. It is designed to study percolation clusters constructed in accordance with the Monte Carlo method using

A. Herega, *The Selected Models of the Mesostructure of Composites*, SpringerBriefs in Physics, https://doi.org/10.1007/978-3-319-89704-2_2

a wide range of algorithms, which allows using it to study various phenomena and processes.

Infinite, or percolation, cluster is a part of the cluster system that connects opposite edges of a solid body or model space in which the growth of clusters is simulated. A percolation cluster is fractal object. Fractal is a basic concept that does not have an exhaustive definition. It was introduced by B. Mandelbrot in the 70s of the twentieth century. The most widespread are three definitions, given by B. Mandelbrot, and not claiming on completeness: (1) an object that is in a sense similar to itself; (2) an object whose fractal dimension is strictly greater than its topological dimension; (3) an object that contains holes of all possible sizes.

Depending on the purpose of the study, the role of percolation clusters in the model is played by various structural elements: internal boundaries (fractal surfaces that separate dissimilar parts of the body; attributive component of composites), the accumulation of a different nature defects, a group of particles, phases of matter, cracks, and others. The basis of its computer implementation is an object-oriented approach, which makes it convenient to vary the objects of research, change their properties and conditions for carrying out computer experiments.

Presented in the book, some implementations of the complex model are intended to solve a range of physical problems of Materials Science of composites in which the dominant role in the formation of properties is played by connected domains and their structure, and therefore, to solve various problems of the classical percolation theory.

2.1   Percolation Clusters of Phases and Interior Boundaries

The construction of a percolation cluster in the model is carried out in accordance with the Monte Carlo method on a square field of more than 130,000 cells. The particles from which clusters are built are in the form of circles. Usually, in model experiments, the field side exceeds particle sizes by 40–50 times. In any experiment, it is necessary to decide on the particle distribution nature: It is either fixed or selected from a discrete normal distribution [11].

The model clusters' growth algorithm assumes a consistent growth of coordination spheres (Fig. 2.1). Control parameters in the algorithm are the number of clustering centers, the relative size of a single particle, the condition for the occurrence of connectivity between elements, as well as the relation between particle sizes forming the clusters. Figure 2.2 shows clusters formed with different values of control parameters.

In the filling of spheres, an essential role is played by the random number generator with uniform distribution. At first, it is used to select the coordinates of the clustering centers, then—that of the centers in which the next act of filling will take place, and, finally, the generator indicates the place in the filled coordination sphere where the next particle of the growing cluster will be located. In the event that the elements' size is not fixed, it is selected from a circle size array (the larger of which is equal

Fig. 2.1 Coordination spheres accumulated on random centers by particles of the same size

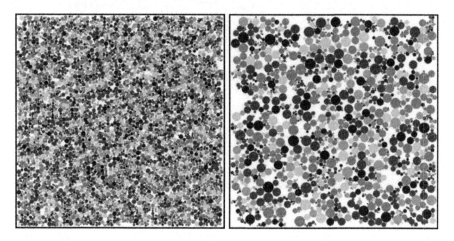

Fig. 2.2 Clusters formed from particles of different sizes, obtained by varying the values of the control parameters

to the diameter from the experiment with identical circles) using a generator with a normal distribution. Generation ceases when a percolation cluster of particles is formed.

In the model, two cluster systems—particles and internal boundaries (or voids)—are actually formed. They are the backdrop for one another as in the mosaics of Maurits Escher (Figs. 2.3 and 2.4). What counts as clusters depend on the study purpose, as well as, for example, the perception of visual images is largely predetermined by the position of an observer.

Fig. 2.3 Model clusters and engraving by M. Escher "Mosaic I"

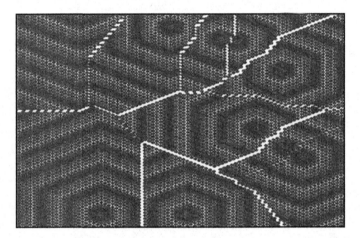

Fig. 2.4 Percolation cluster of internal boundaries

Table 2.1 shows the results of computer experiments [10, 11]. Only quantities whose values make it possible to get an idea of the "physics and geometry" of clusters are presented. The characteristic relative error of calculations is $(10 \pm 3)\%$, and the calculated formulas are described in Appendix B.

According to the data presented in the table, it can be seen that the dispersed composition of particles significantly affects the properties of the material cluster systems. Thus, if particles have an arbitrary diameter (under the condition described above that the maximum value of the diameter of a Gaussian set is equal to the value of a fixed diameter from another implementation of the model), the power of the cluster system of particles is about 1.5 times greater than with a fixed one. The correlation length, in the first approximation, corresponds to the characteristic size of finite clusters of particles, on average, approximately 3 times less in the case of an arbitrary diameter, as well as the radius of gyration (if the mass of the body is considered to be concentrated at one point (reduced mass), located at a distance from

Table 2.1 Parameters of the cluster system of particles and internal boundaries on the surface of the body

Parameters	Fixed diameter	Set of diameters
Power of the cluster system of particles	$(397 \pm 20) \times 10^{-3}$ (5.0%)	$(581 \pm 54) \times 10^{-3}$ (9.3%)
Power of the largest finite cluster of internal boundaries (LFC IB)	$(29.3 \pm 2.8) \times 10^{-3}$ (9.6%)	$(20 \pm 1.8) \times 10^{-3}$ (11.1%)
Power of the percolation cluster of internal boundaries (PC IB)	$(38.4 \pm 3) \times 10^{-3}$ (7.8%)	–
Radius vector of the mass center of the largest finite cluster of particles (LFCP)	286.47 ± 26.82 (9.4%)	315.38 ± 32.06 (10.2%)
Radius vector of the mass center of PC IB	288.14 ± 23.38 (8.1%)	–
Radius of gyration of LFC IB	93.08 ± 9.99 (10.7%)	36.05 ± 3.44 (9.5%)
Radius of gyration of PC IB	116.74 ± 14.03 (12.0%)	–
Average value of the gyration radius of finite clusters of particles (FCP)	84.79 ± 9.65 (11.4%)	25.28 ± 2.61 (10.3%)
Mean correlation length of particle clusters	149.17 ± 13.68 (9.2%)	44.07 ± 4.73 (10.7%)
Relation between the mass of PC IB and the radius of gyration $b = S/R^D$	10.41 ± 1.05 (10.1%)	–
Relation between the mass of LFCP and the radius of gyration $b = S/R^D$	10.13 ± 1.06 (10.5%)	9.95 ± 1.21 (12.2%)
Fractal dimension of PC IB	1.32 ± 0.09 (6.8%)	–
Power growth index of FCP	2.54 ± 0.16 (6.3%)	2.34 ± 0.12 (5.1%)

the axis of rotation equal to the radius of gyration, the moment of inertia will be equal to the moment of inertia of the body), the value of which is particularly important at the stage of material formation.

Dashes in the right column of the table, where it should be the percolation cluster characteristics of internal boundaries, are typical. These parameters were obtained with a relative error exceeding 15%. This is due to the dimensionality of the space in which the percolation "game" occurs: In such a statement of the problem, an infinite (percolation) cluster of boundaries arises extremely rarely, and its absence is typical.

Volumetric task is implemented in a cube containing 10^6 cells and does not need an individual description: The two-dimensional task is easily generalized [10, 11]. Figure 2.5 shows a typical implementation of percolation clusters in a three-dimensional model: on the left—a cluster of balls of arbitrary diameter, on the right—a cluster of voids (internal boundaries)—for clarity in an empty field.

The results of the experiments are presented in Tables 2.2 and 2.3.

Table 2.2 Parameters of percolation clusters of particles and internal boundaries in body volume

Parameters	Fixed diameter	Set of diameters
Power of the cluster system of particles	$(440 \pm 50) \times 10^{-3}$ (11.4%)	$(636.9 \pm 59 \times 10^{-3}$ (9.3%)
Power of PC IB	$(2600 \pm 200) \times 10^{-6}$ (7.7%)	$(800 \pm 60) \times 10^{-6}$ (7.5%)
Radius vector of the mass center of percolation cluster of particular (PCP)	40.73 ± 4.48 (10.9%)	51.57 ± 5.34 (10.4%)
Relation between the mass of PCP and the radius of gyration $b = S/R^D$	12.59 ± 1.35 (10.7%)	6.46 ± 0.65 (10.1%)
Fractal dimension of PC IB	1.49 ± 0.14 (9.4%)	1.39 ± 0.12 (8.6%)
Radius of gyration of PC IB	29.49 ± 3.45 (11.7%)	29.63 ± 1.66 (5.6%)
Degree of anisotropy of PCP A_{xy}	10.64 ± 1.43 (13.4%)	10.02 ± 1.35 (13.5%)
Degree of anisotropy of PCP A_{xz}	20.73 ± 1.71 (8.3%)	26.39 ± 3.25 (12.3%)
Degree of anisotropy of PCP A_{yz}	4.89 ± 0.65 (13.3%)	4.86 ± 0.62 (12.8%)

Table 2.3 Parameters of finite clusters of matter and internal boundaries in body volume

Parameters	Fixed diameter	Set of diameters
Power of LFC IB	$(900 \pm 100) \times 10^{-6}$ (11.1%)	$(400 \pm 40) \times 10^{-6}$ (10.0%)
Radius of gyration of LFC IB	19.43 ± 1.99 (10.2%)	21.28 ± 2.37 (11.1%)
Radius vector of the mass center of LFCP	39.34 ± 4.03 (10.2%)	36.86 ± 3.83 (10.4%)
Relation between the mass of LFCP and the radius of gyration $b = S/R^D$	14.64 ± 1.54 (10.5%)	12.76 ± 1.57 (12.3%)
Degree of anisotropy of LFCP A_{xy}	5.57 ± 0.73 (13.1%)	3.89 ± 0.41 (10.5%)
Degree of anisotropy of LFCP A_{xz}	5.42 ± 0.69 (12.7%)	4.31 ± 0.19 (4.4%)
Degree of anisotropy of LFCP A_{yz}	26.41 ± 1.80 (6.8%)	1.93 ± 0.10 (5.2%)
Mean correlation length of particle clusters	38.58 ± 3.72 (9.6%)	24.39 ± 2.24 (9.2%)
Power growth index of FCP	2.91 ± 0.26 (8.9%)	3.11 ± 0.36 (11.6%)
Average value of the gyration radius of FCP	20.42 ± 2.44 (11.9%)	12.61 ± 1.25 (9.9%)

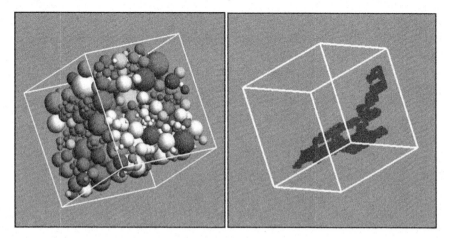

Fig. 2.5 Cluster systems of the volumetric model

2.2 Modeling of Technological Cracks

In the phenomenological model of the evolution of technological cracks in the body volume, the continual percolation problem is solved by the Monte Carlo method [11, 14]. The cluster system is also constructed in a cube of 10^6 cells. Structural elements of the percolation cluster—tracks—are analogous to small technological cracks. The length of tracks in the model varies and takes values 5, 10, 15 length units, and vertical and horizontal components of the deviation angle are selected from the set $\pm(15°, 30°, 45°)$.

Single tracks are considered to be connected if they have at least one common point or the distance between them is equal to some minimum distance (control parameter). The initial coordinate of the track in the cube volume, its length, and orientation angles a random number generator with a uniform distribution defines.

In model experiments, different regimes of cluster formation were studied, and the conditions for obtaining percolation clusters of various structures were determined, and its parameters were calculated.

Results showed that for the parameters of percolation clusters' cracks, the following regularities are characteristic:

- for large tracks of fixed length, an increase in the maximum deviation angle leads to a decrease in the fractal dimension of the percolation cluster, and for small track lengths its growth is observed;
- at a fixed maximum deviation angle of tracks for large angles—as the length increases, the fractal dimension decreases;
- at a fixed track length with the maximum deviation angle increasing, the power of the percolation cluster of the boundaries increases;

- at a fixed value of the maximum angle: at large angles—as the track length increases, the cluster power increases, while at small angles the dependence is inverse;
- at a fixed track length with an increase in the maximum deviation angle, the percolation threshold value decreases;
- for large angles at a fixed value of the maximum angle, the percolation threshold value increases with the track length increasing, for small—vice versa.

2.3 Percolation in the Sierpinski Carpet

Let us consider the continuum percolation problem in the Sierpinski carpet with hybrid branching [8].

The Sierpinski carpet is, as is well known, a two-dimensional analogue of the Cantor set of excluded means [15]. It can be built using a simple algorithm: Each side of the square of a unit area is divided into three equal parts; segments of straight lines passing through the points of division parallel to sides create nine small squares, and the central one is extracted. The procedure is repeated an infinite number of times on each of the eight remaining; the resulting set is a regular fractal with the dimension of self-similarity $D = \ln 8/\ln 3 = 1.892789\ldots$ [15].

Cells of the Sierpinski carpet are considered to be connected if their sides are in contact; in other words, the Sierpinski carpet has an infinite branching (see Appendix C); i.e., the problem of dividing it into parts can be solved by removing an infinite (countable) set of points. Parameters of the percolation transition in the Sierpinski carpet were studied in [7]. In the Sierpinski carpet with hybrid branching, the cells are considered to be connected, if they have a common vertex or if their sides are in contact. It is clear that the modification of rules for the coherency formation leads to a change in percolation parameters of an infinite cluster of carpet cells [8].

According to the described algorithm, we divide any cell of the Sierpinski hybrid carpet of any generation into 9 squares and remove the middle one. Then determine a probability p' of a cell belonging to a percolation cluster in the carpet, i.e., the probability of that it is possible to "flow" over the cell through its constituent squares, each of which is included in the infinite cluster with a probability p.

It is well known that renormalization transformation is a scale transformation for calculation the probability p' of the fact that through a cell it is possible to "leak" over its constituent cells, each of which enters the infinite cluster with probability p. Since the renormalization group transformation [16] should in our case reflect the fact of the coherence presence, a number of suitable combinations in the arrangement of the squares in the cell will be less than the combinatorial one. Given this fact, the renormalization transformation for the hybrid branching carpet has the following form

$$p' = R(p)$$
$$= p^8 + 8p^7(1-p) + 27p^6(1-p)^2$$
$$+ 44p^5(1-p)^3 + 38p^4(1-p)^4 + 8p^3(1-p)^5 \qquad (1.1)$$

with a nontrivial fixed point $p_c = 0.5093$ which determines the percolation threshold [8].

In the first approximation, a correlation length is the characteristic spatial scale of the system; an assessment of the typical size of its constituent clusters. The correlation length index of the percolation system can be found from a relation $v = \ln b / \ln \lambda = 1.801$, where $b = 3$—a number of squares along the side of the cell, $\lambda = (dR/dp)$ $| p = p_c$. The critical exponent of an order parameter is determined from an equality $D = d - \beta/v$, where the Sierpinski carpet dimension serves as an approximation to the dimension D of the percolation cluster; when the space dimension $d = 2$, the value $\beta = 0.193$. Other critical exponents can be determined from the system of equalities of two exponential scaling [4]: an index of the average length of the finite cluster $\gamma = vd - 2\beta = 3.216$; a critical exponent of the heat capacity analog $\alpha = 2 - vd = -1.602$; determining index of the largest size of finite clusters $\Delta = vd - \beta = 1.809$.

2.4 The Development of the Concept of the Immediate Neighborhood at Percolation Models

The immediate neighborhood of the cluster systems' element is one of the key elements of percolation models. Indeed, by modifying the coherency conditions between the clusters' elements (connection at a given distance, when contacting or overlapping), by varying the type of elements (solid, perforated, contour, or fractal), by selecting the type of matrix on which the problem is considered (points, connections, mixed problem; regular, fractal, or random lattice; various continual problems), the type of simulated problems changes crucially [8, 9, 11, 14, 17, 18]. Moreover, the attention that the concept of the immediate neighborhood deserves is due to the need to expand capabilities of computer models in detailing the processes of generation and study of the cluster systems.

The immediate neighborhood, in the first place, is the complex neighborhood; it has the ability to be in one of n states; it is multilayered; it has "crosses of properties"—gradients of the structure in m directions; it is characterized by boundary conditions; it allows the modification of properties and their level.

In the proposed approach, the immediate neighborhood, in fact, is a percolation field itself: one can speak of a distribution of conducting regions in it, ways to determine the entropy of their location, a critical concentration, a classification and other parameters. For the immediate neighborhood, as well as for percolation problems in general, algorithms for determining the measure (in a sense of the dimension theory) in the set of conducting sections, and the dimension describing the scaling behavior of the entropy of their decomposition (see Appendix C); an idea of a relative degree of ordering of the structure is introduced, a suitability of this quantity for estimating a drift of the immediate neighborhood properties is shown (see Appendix D).

References

1. Broadbent, S.R., Hammersly, J.M.: Percolation processes: I. Crystals and mazes. J. Proc. Cambridge Philos. Soc. **53**, 629–641 (1957)
2. Mandelbrot, B.: The Fractal Geometry of Nature. W.H. Freeman and Co., San Francisco (1982)
3. Feder, J.: Fractals. Plenum Press, New York (1988)
4. Sokolov, I.M.: Dimensions and other critical indictors in the percolation theory. Sov. Phys. Usp. **29**, 924–945 (1986). https://doi.org/10.1070/PU1986v029n10ABEH003526
5. Shklovskii, B.I., Efros, A.L.: Electronic Properties of Doped Semiconductors. Springer, Berlin (1984)
6. Trugman, S.A., Weinrib, A.: Percolation with a threshold at zero: a new universality class. Phys. Rev. B **31**, 2974–2980 (1985)
7. Ben-Avraham, D., Havlin, S., Movshovitz, D.: Infinitely ramified fractal lattices and percolation. Philos. Mag. B **50**, 297–306 (1984)
8. Herega, A., Drik, N., Ugol'nikov, A.: Hybrid ramified Sierpinski carpet: percolation transition, critical exponents, and force field. Phys. Usp. **55**(5), 519–521 (2012). https://doi.org/10.3367/UFNe.0182.201205f.0555
9. Herega, A., Sukhanov, V., Vyrovoy, V.: The model of the long-range effect in solids: evolution of structure clusters of interior boundaries, and their statistical descriptors. AIP Conf. Proc. **1909**, 020069 (2017)
10. Herega, A., et al.: Percolation model of composites: fraction clusters and internal boundaries. Int. J. Compos. Mater. **2**, 142–146 (2012)
11. Herega, A.: Physical aspects of self-organization processes in composites. 1. Simulation of percolation clusters of phases and of inner boundaries. Nanomech. Sci. Technol. **4**(2), 119–132 (2013)
12. Herega, A., Vyrovoy, V.: Inter'yernyye granitsy kompozitov: polimasshtabnost' struktury i svoystva silovykh poley (Interior boundaries of composites: the multiscale structure and the properties of force fields). In: Abstracts of the Proceedings of the 4th All-Russian Symposium of the Mechanics of Composite Materials and Structures, Institute of Applied Mechanics RAS, Moscow, 4–6 December 2012
13. Aslanov, A.M., Bekker, M.B., Vyrovoy, V.N., Herega, A.N.: Imitation model of synergetic processes in dynamic disperse systems: Ξ criterion. Tech. Phys. **55**, 147–150 (2010)
14. Herega, A.: Development of the concept of immediate neighborhood at the percolation models of composites. AIP Conf. Proc. **1683**(1–4), 020071 (2015)

15. Bozhokin, S.V., Parshin, D.A.: Fraktaly i mul'tifraktaly (Fractals and Multifractals). R&H Dynamics, Moscow-Izhevsk (2001)
16. Reynolds, P.J., Stanley, H.E., Klein, W.: Large-cell Monte Carlo renormalization group for percolation. Phys. Rev. B **21**, 1223–1245 (1980)
17. Herega, A., Ostapkevich, M.: Computer simulation mesostructure of cluster systems. AIP Conf. Proc. **1623**, 209–212 (2014)
18. Herega, A., Sukhanov, V., Vyrovoy, V.: Multicentric genesis of material structure: development of the percolation model and some applications. AIP Conf. Proc. **1783**, 020072 (2016)

Chapter 3
Percolation Model of the Long-Range Effect

Abstract The second chapter of the monograph describes the percolation model of the set of phenomena observed in crystals during ion implantation and irradiation with light, after which the name «long-range effect» was fixed. The essence of the effect is a change in the material structure at distances from the irradiated surface, which are several orders of magnitude greater than the depth of penetration of irradiation energy. The chapter introduces the concept of the "long-range effect" in irradiated crystals as various manifestations of the second-order phase transition that allows us to interpret the long-range effect as a result of the critical behavior of the amorphized (defective) layer. It is shown that the changes appearing in mechanical fields in the material essentially alter the equilibrium distribution of defects, and these results in the formation of regions with elevated density of defects far from the area of the most probable range of ions. In addition, postulating in the model of long-range character of the stress field makes it possible to estimate the value of the fractal dimension of the percolation cluster in the defect layer. The model also makes it possible to interpret the non-monotonic dependence of the sample properties on the irradiation dose, in particular, microhardness, and also, it allows to explain the presence of «magic» doses at which extremums of properties are observed. The percolation model of the long-range effect in addition to the initial stage of amorphization of crystalline bodies makes it possible to describe the kinetics of accumulation of intermediate substances and final products in main groups of chain processes—polymerization, cracking, oxidation. In model experiments values of the percolation and cluster thresholds, the radius of gyration, the correlation length, the radius of the cluster's center of mass, the degree of anisotropy, and the first five components of the spectrum of Renyi dimensions for percolation clusters constructed using various algorithms are determined. What's more, the possibility of observing the dynamics of the process of cluster system formation is realized.

The set of phenomena observed in crystals during ion implantation and irradiation with light, after which the name «long-range effect» was fixed [1] is studied for about forty years. However, as the authors of [2–4] note, there is no convincing explanation within the framework of classical models of solid body radiation physics, or any complete theory, or unity of views on mechanisms underlying these phenomena.

A. Herega, *The Selected Models of the Mesostructure of Composites*, SpringerBriefs in Physics, https://doi.org/10.1007/978-3-319-89704-2_3

17

The essence of the effect is a change in the material structure at distances from the irradiated surface, which are several orders of magnitude greater than the length of the projective range of ions, i.e., the depth of penetration of irradiation energy. Usually, this results in an increase in the package defects density in the crystal volume in comparison with the region of the ions run (defects of packaging are a violation of spatial periodicity in the alternation of close-packed layers of the crystal lattice) [3], the appearance in the irradiated crystal of anomalous diffusion [5, 6], in the non-monotonic variation (as the radiation dose increases) of the microhardness of a sample at distances up to several hundred micrometers from the defect layer [4, 7], and some others. (In the context of the monography important, that in case of ordinary diffusion, the average value of the squared displacement of the impurity from the initial position is proportional to the time; otherwise, diffusion is anomalous. There is a plethora of reasons for the occurrence of anomalous diffusion; in our book, this is the fractal structure of the percolation cluster of defects).

The chapter introduces the notion of a «long-range effect» in irradiated crystals as various manifestations of the second-order phase transition. A phenomenological percolation model of the phenomenon is proposed, an interpretation of experimental data is given, and a possible mechanism of the effect is discussed.

3.1   The Anomalous Interaction in Semiconductors

A review of even the main experimental results of studying defective structures in implanted or irradiated crystals would require consideration of many dozen works [2, 8]. A few researches describing the characteristic experimental data accumulated by 2010, which have led to the creation of an effect model, are described below.

In research [9], a change in the microhardness (hardness is the property of solids to resist the introduction of another body into it; microhardness is the hardness of the individual phases and structural components of the solids, the hardness of the individual grains, the hardness of the surface layer, etc.) on both sides of a permalloy foil with the thickness of 20 μm was detected when irradiated with halogen lamp of 300 W. The authors found that the dependence of the relative change in the microhardness on the duration of irradiation is non-monotonic, that the interval of irradiation doses is limited from above and from below (the microhardness does not change outside this interval), and that the effect does not depend on the sample heating.

The authors of research [9] explain the effect on the analogy between ion and light irradiation. They believe that in the absence of temperature gradients, and also in the absence of excitation of the material electronic subsystem, «the most probable mechanism is the excitation by ions of elastic (deformation) waves and their interaction with initial structural imperfections». We need to agree with the authors: «the assumption that the generation of deformation waves occurs when light from an incandescent lamp is applied is the most difficult point of the model [9]».

In [7, 10], a change in the microhardness of polycrystalline metals and alloys of several hundreds of micrometers thick was observed at doses of ion irradiation 10^{13} ÷ 10^{16} cm^{-2}. In research, it is shown that the microhardness is a non-monotonic function of the dose with sharp expressed maxima, and moreover, the phenomenon has a threshold (in terms of irradiation energy) character and practically does not depend on the type of ions, the type of metal, the thickness of the foil, the density of the ion current, and arises on both sides of the foil.

The authors of [11] believe that «hypersonic waves are responsible for the long-range effect». They think that «an ion colliding with a surface produces a kind of microexplosion that generates a high-frequency acoustic wave», and, although, «as a rule, the waves quickly damp and by themselves cannot reach the reverse side of the plate, but encountering in their path extended defects they cause their rearrangement, which is accompanied by the emission of secondary waves. Then a kind of chain process arises, which, ultimately, is able to cover the entire thickness of the plate and lead to a change in its properties» [11].

The recent work [3] has substantially supplemented the experimental data on the effect. The authors carried out a metallographic study of silicon samples irradiated with alpha particles with an energy of 27.2 meV at an integrated flux density of $\Phi = 10^{17}$ cm^{-2} with an intensity not exceeding 10^{12} cm^{-2} s^{-1}. The samples were cooled during the irradiation; their temperature was not higher than 100 °C.

At a projection depth of penetration $R_p = 360$ μm, the authors of [3] found not only five layers of defects located below the implant layer (380, 423, 627, 720, 764 μm), but also three at a depth less than the projection (132, 242 and 341 μm). In addition, a small-scale structure of the defect system was observed in the region of R_p. The paper notes that both—the presence of a small-scale structure and the periodicity of the defective system—indicate a wave mechanism for their formation. Herewith, the authors of [3] note that, on the other hand, a well-known autowave mechanism cannot explain the effect of irradiation in the region below the implantation layer since it involves maintaining the amplitude of propagation of the concentration front of radiation defects due to the plastic deformation, which is possible only at a higher sample temperature.

The results of this research, where the defect layers located above the most probable depth of the ionic projective range were experimentally detected, which eliminates the «one-sidedness» of the experimental data, are an essential component in a series of facts that have made it possible to formulate an idea of the long-range effect as about various manifestations of the second-order phase transition.

3.2 Model of the Long-Range Effect. Clusters with Transforming Elements

The perpetual relevance of percolation methods of matter research during the last, approximately, fifty years, has shown the effectiveness of the percolation theory

[12–14] when examining a vast majority of questions relating to the genesis and evolution of connected domains in stochastic processes in a material.

When the elements concentration of some of the material subsystems increases sufficiently and a percolation cluster arises, this leads to a qualitative jump in the body development—to the realization of the structural phase transition. At a moment of the transition, one of the characteristic dimensions of a percolation cluster becomes comparable with the physical body dimensions, and, as a result, the correlation length in the material changes in a jump-like manner, the selected direction appears, the symmetry of the object decreases. Depending on the physical nature of the percolation cluster, this can lead to the appearance of an anomalous diffusion, to hardening effects or to the material destruction, to the appearance of spontaneous magnetization in ferromagnets, to the Mott transition in impurity semiconductors, to changes in the heat and moisture capacity of the body, etc.

The proposed percolation model allows us to interpret the long-range effect as a result of the critical behavior of the amorphized (defective) layer [1].

Assumptions of the model:

- the defect layer located at the depth of the maximum probable projective range of ions is a quasi-flat, discontinuous «cloud» of amorphized regions of various sizes;
- at a critical dose of irradiation, some regions are combined, and a percolation cluster of the amorphized layer appears in the crystal [12–14];
- the field of mechanical stress created by such a layer is long-range, decreasing with distance as r^{-b}, at b, approximately equal to one.

Let us consider the kinetics of stress changes and the possible mechanism of the phenomenon.

At low radiation doses, the defect regions can be considered as separate clusters that create mechanical stresses declining as $\sigma \sim r^{-3}$ [15]. As the dose increases, a disordered region of percolation type emerges thus transforming into an infinite cluster «only under condition that its density exceeds a certain critical value [16]». At this moment, the law, according to which the mechanical stresses decrease with the distance, changes abruptly: They starts to decrease at a considerably lower pace now and can be approximated by the law $\sigma \sim r^{-1}$. As the dose increases further, the percolation cluster of amorphous layer turns into a «continuous wall» with an exponentially decreasing stress field [15]. Naturally, such a drastic character of a change in mechanical fields in the material essentially alters the equilibrium distribution of defects and these results in the formation of regions with elevated density of defects far from the area of the most probable range of ions.

The postulation of the stress field in the long-range nature model makes it possible to estimate the value of the fractal dimension (see Appendix C) of the percolation cluster in the defect layer.

Let us assume that the law of decreasing the value of mechanical stresses with the distance has the form $\sigma \sim r^{-b}$, where $b = 3 - 2D$, D—the dimension of defects. Then for $D = 0$ and $D = 1$. we obtain the known dependences, respectively, for point defects ($\sim r^{-3}$) and linear dislocations ($\sim r^{-1}$) [15], and the fractal dimension of

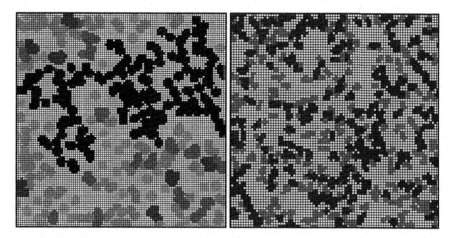

Fig. 3.1 Cluster systems with elements of different sizes and degrees of disorder

the percolation cluster in the defect layer at which the stress field generated by it is long-range will be within $1 < D < 1.5$.

The model also makes it possible to interpret the non-monotonic dependence of the sample properties on the irradiation dose. Indeed, as the dose increases, the percolation cluster, which consists of regions of some disorder degree, loses connectivity and is replaced by a cluster of more disordered regions. During the irradiation, this occurs repeatedly and leads to a fluctuation in the value of the mechanical stresses, to a well-defined non-monotonicity of the irradiated sample properties, in particular, microhardness, and also, it allows to explain the presence of «magic» doses at which extremums of properties are observed.

3.3 Computer Implementation of the Model

A continuous percolation model of amorphous layer formation was studied in computer realization by the Monte Carlo method. A degree of disorder in the defect regions in the model can be varied that corresponds to a change in the energy of implanted ions (Fig. 3.1). Regions are considered to be connected in the case of contact; the degree of disorder increases if they overlap. Four degrees of disorder are provided, as well as the modes that enable any combination of types of modified regions; i.e., there is a possibility to track the formation of the percolation cluster out of all possible combinations of small clusters, which are differently disordered.

The evolution of regions of different degrees of amorphization as the model dose increases is depicted in Fig. 3.2. This can be seen, just as is expected, that the initial growth is replaced by a decrease: The formation of more disordered regions takes place at the expense of those, which are more ordered. One can observe changes

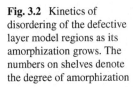

Fig. 3.2 Kinetics of disordering of the defective layer model regions as its amorphization grows. The numbers on shelves denote the degree of amorphization

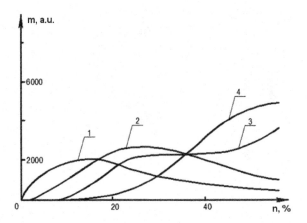

in the character of forming regions with various amorphization degree in the model experiments.

For example, it can be seen in Fig. 3.2 that at the final stage of the percolation cluster formation in the defective layer, the speed of generation of model regions with fourth degree of disorder decreases; i.e., the process of growth of these regions is delayed while that of the third degree accelerates.

In model experiments values of the percolation and cluster thresholds, the radius of gyration, the correlation length, the radius of the cluster's center of mass, the degree of anisotropy, and the first five components of the spectrum of Renyi dimensions (see Appendix C) for percolation clusters constructed using various algorithms, are determined. What's more, the possibility of observing the dynamics of the process of cluster system formation is realized.

3.4 Long-Range in Construction Composites

Creation of building composite materials with desired ranges of properties is a complex multifactor process. This is due to the multicomponent composition of the material and the range of characteristic sizes of components having an extent of three orders—from tens of microns (grain of cement) to tens of millimeters (large aggregate).

It is well known that the strength of structural expanded clay concrete is significantly affected by a grain composition (a characteristic distribution of material fractions over the sizes), an aggregate porosity, a water–cement ratio (the ratio of water mass to cement mass in the process of making concrete. The impermeability of concrete depends on its value), an aggregate-structural factor (a parameter that describes the mechanical properties of a mixture by its microstructure) [17, 18]. In literary sources, the importance of the strength and coarseness of expanded clay fraction for the production of structural expanded clay concrete was repeatedly pointed

Table 3.1 Dependence of the strength of expanded clay concrete on the fraction size, strength, and volume density of expanded clay

Fraction size, mm	Strength of expanded clay, MPa	Volume density of expanded clay	Average density of concrete, kg/m^3	Compression strength, MPa		
				1 day	7 days	28 days
0–10	5.93	0.54	1572	27.3	31.5	34.5
0–10	5.93	0.55	1572	25.5	29.1	37.6
0–10	5.93	0.60	1467	26.5	29.9	32..3
0–10	5.93	0.70	1312	22.3	23.6	29.1
0–10	5.93	0.71	1330	20.7	25.7	30.3
5–10	4.4	0.42	1486	23.8	25.8	31.4
5–10	4.4	0.43	1481	22.1	22.8	30.4
10–20	3.8	0.45	1454	19.7	19.8	30.7
10–20	3.8	0.42	1527	19.2	23.8	29.2

out [17, 18]. The data of numerous experiments showed that with decreasing of the coarseness of expanded clay fraction, the strength of lightweight concrete increases, since small granules have higher strength.

When designing the composition of expanded clay concrete, it is also necessary to take into account the influence of the volume density of expanded clay on its properties. In [19], experimental studies of the influence of the concentration of expanded clay on the strength of expanded clay concrete (Table 3.1) are described.

From the analysis of the data presented in the table, one can see:

(1) the dependence of the strength of expanded clay concrete on the volume concentration of expanded clay has a nonlinear character: It increases with the growth of the concentration to 0.55 and decreases with its further growth to 0.71;
(2) the optimum concentration of expanded clay aggregate in concrete is $0.54 \div 0.6$;
(3) expanded clay concrete with the inclusion of expanded clay fraction from zero to 10 mm has the greatest strength: the strength increases by approximately 20% with an increase in the average density by $(3 \div 5)\%$.

Complementary interpretations of these results are possible on the basis of ideas of building Material Science and the percolation theory.

The first interpretation considers the relationship between volume fractions of expanded clay, sand, cement, and water. As the content of expanded clay in concrete increases, the strength of the latter increases due to the aggregate strength. If the content of expanded clay is too high, the probability of destruction of concrete over the grain of aggregate, not the mortar part, increases.

The second is based on a qualitative analysis of the properties of the percolation cluster (quasicluster) of expanded clay granules that occurs in concrete when the volume density of the aggregate reaches a critical value. If we proceed from the fact that mechanical stresses in concrete during hardening are due to an increase in the volume of the mortar part as a result of hydration processes (the process of joining water and cements molecules, resulting in a plastic mass, leads to the formation of concrete) and, consequently, all-round compression of expanded clay granules, then at a critical concentration of expanded clay the value of mechanical stresses in the material will abruptly increase in accordance with the mechanism described in the previous section. Percolation theory interprets this fact as the result of a structural phase transition, which provokes the increase of stresses in a system. This is due to the emergence in the material system of a selected direction and extended formations that create mechanical stresses which fall off much more slowly than at undercritical concentrations. It is these internal stresses that increase the material strength.

A further increase in the concentration of granules leads, in accordance with the provisions of the percolation model, to a decrease in strength [1, 15]. The presence of such a maximum in the dependence of the strength on the concentration makes it possible to draw an analogy with the effect of long-range interaction in semiconductors and metals.

Speaking about obtaining samples of greater strength when using a fraction $(0 \div 10)$ mm. This result requires further study. It is possible that this is an artifact associated with the samples dimensions $(15 \times 15 \times 15)$ cm^3, i.e., the fact that in such a small volume it is on the basis of the fraction of the smallest sizes that the emergence of a formed percocluster may occur.

3.5   Conclusion

The percolation model of the long-range effect in addition to the initial stage of amorphization of crystalline bodies makes it possible to describe the kinetics of accumulation of intermediate substances and final products in main groups of chain processes—polymerization, cracking, oxidation, as well as to facilitate the solution of problems of establishing connections between the structure of matter and kinetic constants characterizing the reactive capacity.

If the effect is a universal property of solids, it should also appear in polymers, ceramics, composites of various nature and other materials. The solution of this question is seen on the path of extensive experimental research and the derivation of a detailed theory of the phenomenon.

The model admits the presence of similar phenomena in quasi-two-dimensional objects. This can lead to the creation of functional coatings, in particular, to the cladding the surfaces of products that will find application in Microelectronics, Mechanical Engineering, Construction, and other industries.

The authors of [16] studied the mechanical properties of high-strength steel doped with boron. Interfacial segregation of nitrides and oxides of boron has been detected, and it is shown that boron is separated from the solid solution in interphase boundaries at temperatures exceeding (430 ± 20) °C. A correlation between parameters of the interface structure and mechanical properties of the samples was found.

As known, a perovskite is a yellow, brown, or black magmatic mineral containing rare earth element, mostly of calcium titanate. The interest to manganese-lanthanum perovskites is related to magnetoresistance effect (dependence of the electrical resistance of a body on an external magnetic field) observed near the phase transition [17], as well as to the ability to manufacture light-emitted diodes and to create photovoltaic elements in which electricity is converted into visible light in a range not available for other types of photovoltaic elements. In [18], the regularities of the effect of temperature on the structure, magnetic and resistive properties of materials obtained by sol-gel technology were studied. (As known, gel is a semisolid colloidal suspension of a solid dispersed in a liquid, and sol is a fluid suspension of a colloidal solid in a liquid.)

The magnetoresistive effects are observed in manganese-lanthanum perovskites at low temperature. And, as a rule, it is explained by tunneling via mesoscopic boundaries of particles. In the research, it is shown that the optimum sintering temperature is 700 °C, at which the increase in the magnetoresistive effect is $(15 \div 25)\%$ [18].

In works [19, 20], it was specified that, on the contrary to the established idea, the electrical activity of internal boundaries in silicon, as well as in silicon bicrystals (a crystal consisting of two single crystals, the contact surface of which represents a high-angle boundary; high-angle (or large-angle) boundary is the boundary, on both sides of which the disorientation of the crystal lattices is more than 10–15°. Large-angle boundaries exist in crystal lattices more often than small-angle boundaries, since the embryos of neighboring grains are usually very different in space. Usually, the large-angle borders have a looser structure and migrate much faster.) produced by solid-state bonding method with background (oxygen, carbon, aluminum) and dopant (boron) impurities is induced by dangling bonds that emerge as a result of the discrepancy between lattices of neighboring grains. It is established that for the material a high density of structural defects and grain boundaries is characteristic. Due to the small grain size, a particular defect of the grain boundary topology, the so-called triple junction, takes a dominant role for grain growth and atomic transport [21]. The role of this defect was experimentally confirmed in tomographic studies of nanocrystalline copper. The authors of [21] demonstrated that triple junctions have 100–300 times higher diffusion coefficient of Ni than standard grain boundaries. According to the authors of [21], this is due to considerably lower activation energy.

In [22], a technique for influencing on the microstructure of thin lanthanum nickel films (lanthanum nickel oxide $LaNiO_3$) in the sol-gel synthesis process was tested. The temperature dependence of the resistivity and the magnetoresistance were studied, and the influence of grain boundaries on the transport properties of electrons was investigated. The results show that the grain boundaries have a significant influence on the transport behavior of the electrons in lanthanum nickel oxide films at a low-temperature region, which can be captured by a variable-range hopping model.

(Hopping conductivity is a low-temperature conduction mechanism in semiconductors. It is realized due to hopping of charge carriers between different localized states.) The possible increase of metal-insulator transition temperature is ascribed to Anderson's localization. (It is a phenomenon that arises when waves propagate in a medium with spatial inhomogencities. The meaning of the concept is that multiple scattering of waves by inhomogeneities and their interference leads to the impossibility of propagating traveling waves, and the appearance of a standing wave.) The centers of localization, according to the authors of [22], are grain boundaries. In this case, boundaries show a minor effect on the transport behavior, but elevate the residual resistivity.

The authors of [23] carried out measurements of the electrical conductivity of thin films of α-sexithiophene crystallites by atomic force microscopy technique. It was clearly shown that the grain boundaries are the dominant factor influencing on the electrical properties of organic thin films. Moreover, it is established that the hole transport across the grain boundary is significantly affected by doping of the material with positive oxygen ions.

Models and concepts: In 2007, the authors of [24] pointed to the lack of polycrystalline composites models, in which the explanation of properties was reduced to the analysis of granular texture and statistical parameters of crystal lattices; and in which grain sizes were not taken into account, the presence and parameters of intergranular boundaries were not taken into account. Further studies confirmed that these are essential components that are necessary for the quantitative evaluation and forecasting of strength, plasticity, transport, and other characteristics of composite materials.

In [25], the fundamental physical principles of mesomechanics of surface layers and internal interfaces in loaded structurally inhomogeneous media are discussed. Surface layers and internal interfaces are classified as autonomous mesoscopic structural levels of deformation, which play an important functional role in the mechanical behavior of solids in various fields of external influences.

In [26], the results of numerical modeling of residual thermal stresses and strains which develop during the cooling of the inhomogeneous material are presented. It is shown that thermal stresses and strains are localized in thin near-surface layers and in the volume of material along the interfaces of structural elements. The concentration of residual stresses leads to a decrease in the external load, which causes a local flow in the volume of the material, and the emergence of areas of the localization of volumetric plastic strains in thin surface layers means the emergence of destruction seats. This result partly explains why destruction almost always starts from the material surface.

Based on the molecular dynamic modeling of copper crystallite stretching, it was shown in [27] that the generation of defects is initiated at the boundary of regions with a disproportional strain. The initial stage of plastic deformation initiation caused by the formation of nanoscopic-size blocks was investigated in this research.

In [28], computer modeling techniques were used for studying the features of deformation and destruction of materials with a large share of interfaces («interfacial materials») under the influence of complex alternating loads. The effect of the frequency of cyclic influence on the nature of the destruction of samples, their deformability, the ability to dissipate the loading energy, as well as the change in the interface material response with a variation in the volume share of interfaces were analyzed. It was shown that high-frequency vibrational influences with frequencies exceeding their own can significantly increase the deformation capacity and dissipative characteristics of such materials.

In [29], the problems of modeling of processes of deformation and destruction of materials with a complex internal structure at different scale levels were considered from the standpoint of physical mesomechanics. The author of [29] showed that at high degrees of straining, substructures of mesoscopic scales make an important contribution to the flow resistance. It was shown that the formation of a blocks' hierarchy (structural elements of various sizes strain) depending on the initial inhomogeneity and loading conditions is a characteristic feature of the strain development at the mesolevel.

A multilevel model of the strained polycrystal, according to which self-coupling of distorting grains is based on the development in their boundaries of rotational-wave flows of planar structural transformations, was developed in [30]. Computer modeling showed that there are two types of rotational-wave flows, which are determined by the angle of disorientation of adjacent grains. The first type of grain boundary flows develops in small-angle boundaries and is characterized by a small curvature. Such flows generate dislocations to the grains' volume. The second-type flows are characteristic for large-angle boundaries; they form in frontier zones bands of curvature, which cause fragmentation of grains, and significantly violate the translational invariance.

In [31], a molecular dynamic modeling of diffusion processes in the intergranular boundary region was carried out. It was shown that in the grain boundary region, high-temperature heating leads to significant atomic displacements, determining the active grain boundary diffusion. From the analysis of calculation results, it follows that parameters of the grain boundary diffusion can be estimated with sufficient accuracy on the basis of molecular dynamic calculations. This makes it possible to investigate atomic mechanisms of the influence of the temperature and external mechanical influences on diffusion processes and structural rearrangements both in the intergrain region and in the grain body.

It was shown in [32–34] that the growth of the relative share of interphase layers is the structural basis for increasing the strength of elastomers with a fine aggregate. As known, elastomers are highly elastic polymeric materials that restore the original shape and size after deformation; a typical elastomer is rubber. The influence of interphase interactions on physical-mechanical properties of polymeric dispersed-filled composite media was discussed. Relations that establish the interrelationship between macromechanical characteristics of a heterogeneous medium and fractal parameters of interphase regions are proposed.

4.2   Fractal Nature of Interior Boundaries' Networks. Force Field Model of Polyscale Network

Disruptions in the homogeneity of the material spontaneously arise in different parts of the physical body. When the degree of heterogeneity of the material is sufficiently increased, and collective effects appear in the interaction of structural elements, a qualitatively new stage of its evolution arises.

Orientations of linear inhomogeneities, appearing at an early stage of material formation, are not random, of course, but not correlated as well. When their density grows, prevailing directions appear in the orientation of newly generated linear defects. This is due to the fact that the fields of deformation depend substantially on the configuration of inhomogeneities. In the case of quasilinear inner boundaries, the values of deformation tensor components are proportional to r^{-1} [35], and, therefore, their action may be substantial at comparatively large distances that several times exceed interatomic spacing.

Thus, the fields of deformation of linear inhomogeneities activate locally the generation of parallel defects and, to a smaller degree, prevent perpendicular growth of defects. Consequently, a typical spatial feature appears, a self-affine multifractal (see Appendix C) «pattern» of the internal boundaries (Fig. 4.1), very simple analogs of which in the plane and the volume, respectively, can be modified fractals, such as the Sierpinski carpet (Fig. 4.2), Menger sponge, and their supplements (Fig. 4.3) [36, 37].

The proposed model of the force field is based on the statistical self-similarity of the boundary networks and modifications of the Sierpinski prefractals. The model is constructed on the assumption that any network of internal boundaries can be obtained with a predetermined accuracy by imposing the required number of modifications of regular prefractals of random generations, similar to any practically significant function that can be expanded into a Fourier series. Moreover, modeling by the Sierpinski carpets allows taking into consideration a substantial aspect of the process, such as multiplicity of interacting spots of generation of new structural units.

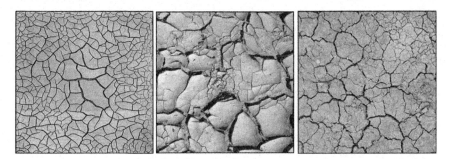

Fig. 4.1 Surface cracks

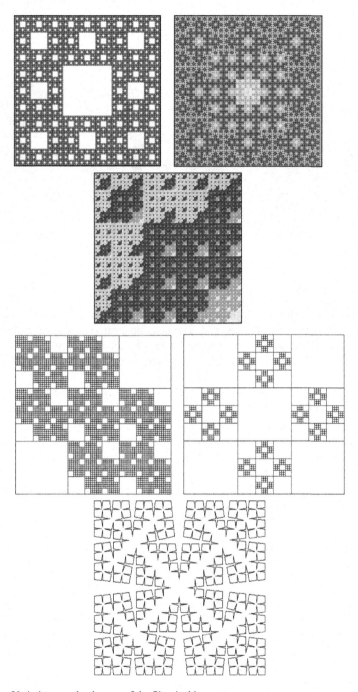

Fig. 4.2 «Variations on the theme» of the Sierpinski carpet

Fig. 4.3 The Menger sponge and its supplement in the second step of the construction [36]

We assume that straight sections of the Sierpinski square correspond to quasilin-ear internal boundaries of the material. Then, we determine analytically force fields generated by multiscale networks of the internal boundaries of the Sierpinski pre-fractals of three types, which differ in symmetry, on the random step of partition m [4, 6, 38].

We consider a «wire» model of the Sierpinski carpet proposed in [38]. Let the orig-inal square frame be divided by four «wires» into nine equal squares. This procedure is repeated several times for each of 8^m frames obtained at every next step (except for the central ones). Let also each generated frame of any «generation» with the linear density λ contain point sources, generating the fields with the intensity $E \sim 1/r^2$.

Let the Sierpinski carpet with the length of the generating square side equal to $2H$ be arranged so that its center coincides with the origin of coordinates, whereas the sides are parallel to the axes. The components of the vector of intensity, generated by the section, are defined by the relations

$$\begin{cases} E_x = \lambda \, (\sin \alpha_2 - \sin \alpha_1)/r \\ E_y = \lambda \, (\cos \alpha_2 - \cos \alpha_1)/r \end{cases} \tag{4.1}$$

where α_i is the angle between the perpendicular of the length r, dropped from the point where the intensity is determined to the section or its continuation, and the corresponding direction to the end points of the section.

We use the designations

$$A(u, v) = (u^2 + v^2)^{-1/2}, \; B(u, v) = v/[u(u^2 + v^2)^{1/2}], \tag{4.2}$$

$$\xi(n, p) = -\xi + (-1)^n p, \, \eta(n, p) = -\eta + (-1)^n p, \tag{4.3}$$

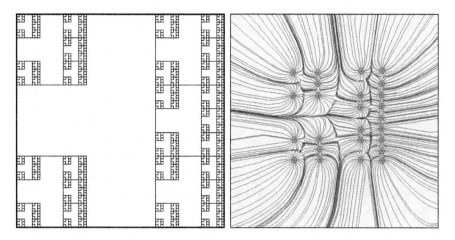

Fig. 4.4 Modification of the Sierpinski carpet and its force field

then the component of the vector of intensity E, generated by the m-th generation carpet at random points, not lying on the straight lines in the network sections, can be calculated from the recurrent relations:

$$E_x = X_m(\xi, \eta) = \sum_{i=1}^{2} \{X_{m-1}(\xi, \eta(i, H'')) + X_{m-1}(\xi(i, H''), \eta)\} + \sum_{j=1}^{2} [X_{m-1}(\xi(j, H''), y(i, H''))$$

$$+ \lambda(-1)^j (A(\xi(j, H'), \eta(i, H')) - A(\xi(j, H), \eta(i, H')) + B(\xi(i, H'), \eta(j, H))$$

$$- B(\xi(i, H'), \eta(j, H')))]\}; \tag{4.4}$$

$$E_y = Y_m(\xi, \eta) = X_m(\eta, \xi), \tag{4.5}$$

where

$$H' = H/3, H'' = 2H/3; \tag{4.6}$$

$$X_0(\xi, \eta) = \lambda \sum_{j=1}^{2} \sum_{i=1}^{2} [(-1)^i A(\xi(i, H), \eta(j, H))$$

$$+ (-1)^{j+1} B(\xi(i, H), \eta(j, H))]; \tag{4.7}$$

$$Y_0(\xi, \eta) = X_0(\eta, \xi)$$

$$= \lambda \sum_{j=1}^{2} \sum_{i=1}^{2} \left[[(-1)^i A(\eta(i, H),), \xi(j, H))\right.$$

$$\left. + (-1)^{j+1} B(\eta(i, H), \xi(j, H))\right]. \tag{4.8}$$

We consider an analog of the Sierpinski carpet with one second-order axis of symmetry (Fig. 4.4).

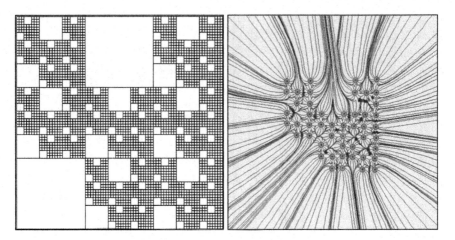

Fig. 4.5 A non-symmetrical modification of the Sierpinski carpet

Using the same designations and taking into account $h = H/3^m$, the field for such carpet is determined from the relations:

$$X_n(\xi; \eta) = X_{n-1}(\xi - 2h_{n-1}; \eta) + \sum_{i=1}^{2}\{(-1)^{i+1} B(\xi + h_n; \eta(i, h_{n-1}))$$

$$+ \sum_{j=1}^{2}[X_{n-1}(\xi(i, 2h_{n-1}); \eta(j, h_{n-1}))$$

$$+ (-1)^i\{A(\xi(i, h_{n-1}); \eta(j, h_n))$$

$$- A(\xi(i, h_{n-1}) - 2h_{n-1}; \eta(j, h_{n-1}))\}]\}; \tag{4.9}$$

$$X_n(\xi; \eta) = X_{n-1}(\xi - 2h_{n-1}; \eta) + \sum_{i=1}^{2}\{(-1)^{i+1} B(\xi + h_n; \eta(i, h_{n-1}))$$

$$+ \sum_{j=1}^{2}[X_{n-1}(\xi(i, 2h_{n-1}); \eta(j, h_{n-1}))$$

$$+ (-1)^i\{A(\xi(i, h_{n-1}); \eta(j, h_n))$$

$$- A(\xi(i, h_{n-1}) - 2h_{n-1}; \eta(j, h_{n-1}))\}]\}; \tag{4.10}$$

We define analytically the force field created by the multiscale network of internal boundaries, the modification of the Sierpinski square, which does not have axes of symmetry (Fig. 4.5). It can be calculated from the recurrent relations:

$$X_1(\xi;\eta) = \sum_{i=1}^{2}\sum_{j=1}^{2}\sum_{l=1}^{2}(-1)^i[A(\xi(i,h_1);(\eta(j,2l-1)h_1)/3))$$
$$- B(\xi(j,(2l-1)h_1/3;\eta(i,h_1))];\tag{4.11}$$

$$Y_1(\xi;\eta) = X(\eta;\xi)\tag{4.12}$$

$$X_n(\xi;\eta) = X_{n-1}(\xi+2h_{n-1};\eta-2h_{n-1})$$
$$+ \sum_{i=1}^{2}\{X_{n-1}(\xi;\eta(i,h_{n-1})-h_{n-1})$$
$$+ (-1)^i[A(\xi-h_n;\eta(i,2h_{n-1})-h_{n-1})$$
$$- B(\xi(i,2h_{n-1})-h_{n-1};\eta-h_n)]$$
$$+ \sum_{j=1}^{2}X_{n-1}(\xi(i,2h_{n-1});\eta(j,h_{n-1})+h_{n-1})$$
$$+ (-1)^i\{A(\xi(i,h_{n-1});\eta(j,h_{n-1})+2h_{n-1})$$
$$- A(\xi(i,2h_{n-1})+h_{n-1};\eta(i,h_{n-1})$$
$$+ ((-1)^j-1)h_{n-1})+B(\xi(i,h_{n-1})$$
$$+ ((-1)^j-1)h_{n-1};\eta(i,h_{n-1}))\}]\};\tag{4.13}$$

$$Y_n(\xi;\eta) = Y_{n-1}(\xi+2h_{n-1};\eta-2h_{n-1})$$
$$+ \sum_{i=1}^{2}\{X_{n-1}(\xi;\eta(i,h_{n-1})-h_{n-1})$$
$$+ (-1)^i[A(\xi(i,2h_{n-1})-h_{n-1};\eta-h_n)$$
$$- B(\eta(i,2h_{n-1})-h_{n-1};\xi-h_n)]$$
$$+ \sum_{j=1}^{2}[Y_{n-1}(\xi(i,2h_{n-1});\eta(j,h_{n-1})$$
$$+ h_{n-1})+(-1)^{i+1}\{B(\eta(j,h_{n-1})+2h_{n-1};\xi(i,h_{n-1}))$$
$$- B(\eta(i,h_{n-1})+((-1)^j-1)h_{n-1};\xi(i,2h_{n-1})$$
$$+ h_{n-1})+A(\xi(i,h_{n-1})$$
$$+ ((-1)^j-1)h_{n-1};\eta(i,h_{n-1}))\}]\};$$
$$n = 2,3\ldots m;\tag{4.14}$$

$$E_x = E_x(x;y) = X_m(-x;-y);$$
$$E_y = E_y(x;y) = Y_m(-x;-y);$$
$$h_n = H\cdot 3^{n-m};\tag{4.15}$$

4.3   The Oscillatory Interaction of Different Scales Structures. Stochastic Model of the Interior Boundaries' Interaction

Solid-state oscillatory reaction: Statistical self-similarity in the arrangement of internal boundaries leads to a certain «multiplicity» in the configuration of the fields, and the coherent interaction between commensurate elements of the structure—inhomogeneities of one scale level—to the appearance in local areas of a physical body as a result of interference of energy prerequisites for the formation of larger boundaries, which play the role of initial elements for a structure of a higher level [6, 10, 12]. In turn, the fields of the deformation of larger boundaries, affecting smaller boundaries, provoke their further development. This occurs in synchronism in all scales and is, in fact, an oscillatory interaction between the internal boundaries of different scales [10]. This provokes the structure inflation, and, thus, contributes to the growth of the heterogeneity of the material texture, leads to the emergence of hierarchical cosubordination, and, as a consequence, to the generation of fundamentally different structures.

Let us consider these processes as a rapidly damping solid-state oscillatory reaction.

Let the physical body represent an oscillatory system in the sense indicated. It is an autonomous distributed non-conservative system with damped anharmonic oscillations. Assuming that the system has generalized force of resistance, the proportional rate of energy distribution between quasilinear boundaries of different scaling levels, then with a correction to the level of legitimacy of the assumptions made, the equation of motion will have the standard form:

$$x'' + 2\gamma x' + \omega^2 x = 0, \tag{4.16}$$

where γ is the generalized damping coefficient and ω is the cyclic frequency.

Now, we estimate the conventional period T of such oscillations [39] for a system of quasilinear boundaries in a rectangular parallelepiped. We introduce a quantity inverse to the coefficient of body stiffness, i.e., the compliance $C = 1/k$, which determines how far the process of cracking may propagate in scales and which has the following calculation formula:

$$C = 8a^3/(Eh^3b), \tag{4.17}$$

where a is the length of crack or quasilinear boundary, h is the distance from it to the body edge, b is the plate thickness, and E is the Young modulus [40]. Assuming γ as a small quantity, we have $\omega^2 \approx \omega_0^2 = 1/(m \cdot C)$; then, we have

$$T = 4\pi (2a^3 m/Eh^3 b)^{1/2}, \tag{4.18}$$

where m is the mass of then body.

Another estimation of the conventional period can be obtained by determining the logarithmic decrement of damping in terms of consecutive (with the period interval) values of the system energy Wn [39, 41],

$$T = (1/\gamma) \cdot (W_n - W_{n+1})/(W_n + W_{n+1}). \tag{4.19}$$

Interior boundaries' interaction: In research [42], a mathematical model of the evolution of abstract systems that consist of parts interacting according to random laws was presented. In [10, 38], it is used to describe the interaction of internal boundaries of composites.

As known, the dynamic system is a mathematical object corresponding to real systems, the evolution of which is uniquely determined by the initial state. The dynamic system is defined by a system of equations that admit the existence of a unique solution for each initial condition on an infinite time interval. Let the material structure be an open nonlinear dynamic system with the interacting internal boundaries of three scaling levels, and assume that the evolution of the energy state of boundaries is described by a system of bilinear iteration equations

$$\Phi_1(x, y, z) = \begin{cases} x_{n+1} = x_n - k_{xy}px_n^2 + k_{yx}qy_n^2 + x_{in} \\ y_{n+1} = y_n + k_{xy}px_n^2 - (k_{xy} + k_{yz})qy_n^2 + k_{zy}rz_n^2 \\ z_{n+1} = z_n + k_{yz}qy_n^2 - (k_{zy} + k_{out})rz_n^2 \end{cases} \tag{4.20}$$

where x, y, and z are dynamic variables that determine the potential energy of the internal boundaries of a specific scaling level and x_{in} is the energy of an external effect. The coefficients k_{ij} specify the share of energy passing between the boundaries of different scales, and the coefficients p, q, and r specify the share of energy used for reconstruction, with $\{k_{ij}\}$ and $\{p, q, r\} \in (0,1)$, $\{x, y, z\} \in R$.

The system of equations has a steady-state solution obtained analytically

$$x_{st} = \sqrt{\frac{x_{in}}{k_{xy}p}\left(1 + \frac{k_{yx}}{k_{yz}}\left(1 + \frac{k_{zy}}{k_{out}}\right)\right)};$$

$$y_{st} = \sqrt{\frac{x_{in}}{k_{yz}q}\left(1 + \frac{k_{zy}}{k_{out}}\right)};$$

$$z_{st} = \sqrt{\frac{x_{in}}{k_{out}r}} \tag{4.21}$$

and is not integrated in a general form.

As x_{in} increases, two options of system evolution are possible. In the first option, a cascade of period doubling bifurcations appears and the Feigenbaum scenario [43, 44] of transition to chaos is implemented (Fig. 4.6). (The Feigenbaum scenario (period doubling scenario) is one of the typical scenarios for the transition from order to chaos, from a simple periodic mode to a complex aperiodic one with an infinite

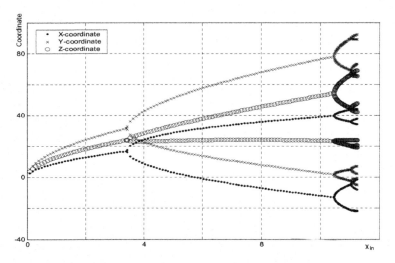

Fig. 4.6 Bifurcation diagram scenario of doubling period for the system with $k_{xy} = 0.5$, $k_{yx} = 0.4$, $k_{yz} = 0.3$, $k_{zy} = 0.3$, $k_{out} = 0.4$, $p = 0.08$, $q = 0.02$, $r = 0.015$. The first three bifurcations are shown

period doubling.) In this case, attractors comprising 2^n points are observed in the phase space (Fig. 4.7). (Attractor (from *English attract*) is a compact subset of the phase space, possessing a certain set of properties, one of which is the presence of the attraction region, i.e., of such set of points that trajectories of dynamic systems emerging from them tend to the attractor on infinitely large intervals of time.)

Let us define still several terms of the theory of dynamical systems. Phase space is an abstract space associated with a specific dynamic system, the points in which uniquely characterize all possible states of the given system. Phase trajectory is a curve in the phase space, made up of points representing the state of the dynamic system at consecutive moments of time during the entire evolution time. And a limit cycle is the simplest regular regime in the phase space, which describes the periodic change of the dynamic state of the system; has the form of an isolated closed trajectory.

Quasiperiodic regime is a regular regime of the dynamic system motion in phase space, characterized by two frequencies. The image of the regime in the phase space is a torus. The fundamental frequency of such a regime is determined by the motion of the image point along the trajectory of the center of the generatrix of the torus circle (i.e., by the limit cycle), and the second frequency is determined by the rotation along the generatrix of the torus. Synchronization of oscillations is the coincidence of frequencies.

In the second case, the periodic mode is followed by a situation similar to the Hopf bifurcation [44, 45] (bifurcation is the situation when the solution of any equation undergoes a qualitative change when a certain parameter reaches a critical value; it is of fundamental importance for the study of dynamic systems. The Hopf bifurcation is

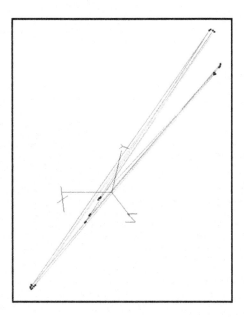

Fig. 4.7 Attractor consisting of 32 points which occurs when parameters are $k_{xy} = 0.5$, $k_{yx} = 0.4$, $k_{yz} = 0.3$, $k_{zy} = 0.3$, $k_{out} = 0.4$, $p = 0.05$, $q = 0.085$, $r = 0.04$, $x_{in} = 6.44$

the phenomenon when a limit cycle emerges from the point, when a certain parameter reaches a critical value), which, as is known, leads to a quasiperiodic mode and to the appearance in the phase space of an attractor in the form of two closed lines (Fig. 4.8). Further increase in x_{in} leads to bifurcation resulting in a chaotic mode, leading to the appearance of a strange attractor in the phase space (Fig. 4.9). Strange attractor is an attracting set of unstable trajectories in the phase space of the dissipative dynamic system; in contrast to the attractor, it is not a manifold (i.e., a curve or a surface); has a fractal structure. When a quasiperiodic regime is realized, it is possible that a synchronization of the model may occur. In such a situation, a periodic regime in the system occurs with a period depending on the level of synchronization (Fig. 4.10).

Another example of the evolution of a system of equations for other coefficients is shown in Fig. 4.11. It is interesting, in particular, by the fact that the fine structure of the strange attractor is visible.

Besides the aforementioned scenarios of chaos development in the system, it is possible to observe their various combinations.

Chaotic regime is a phenomenon in the theory of the dynamic systems, in which the system behavior looks as if it is random, despite the fact that it is determined by deterministic equations. The reason for chaos appearance is the sensitivity of the system to the initial conditions: Infinitely closely spaced points of the trajectory diverge over a long distance with time, which leads to unpredictability of the system behavior over long times. Further increase of x_{in}, when the system is in the determinate chaotic regime, leads to a change of shape of the chaotic attractor. In such a case, we

Fig. 4.8 Attractor occurring
at values of parameters: k_{xy}
$= 0.5, k_{yx} = 0.4, k_{yz} = 0.3,$
$k_{zy} = 0.3, k_{out} = 0.4, p$
$= 0.065, q = 0.03, r = 0.03,$
$x_{in} = 6.05$

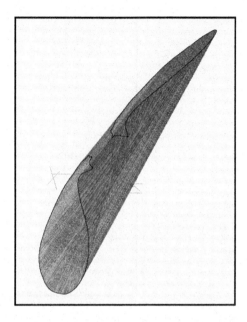

Fig. 4.9 Strange attractor
occurring at values of
parameters: $k_{xy} = 0.5, k_{yx}$
$= 0.4, k_{yz} = 0.3, k_{zy} = 0.3,$
$k_{out} = 0.4, p = 0.008, q$
$= 0.005, r = 0.0057, x_{in}$
$= 39.65$

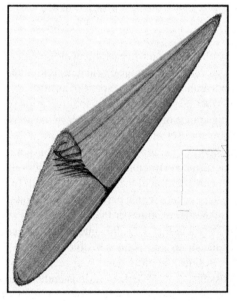

observed the intermittency; that is, chaotic behavior of the system on some intervals
changes to periodic. In Fig. 4.12, some compression of line segments connecting
points of the phase trajectory corresponds to the above alternation. In such a case,
we observed multiple attractors and hysteresis with respect to the parameter x_{in}.

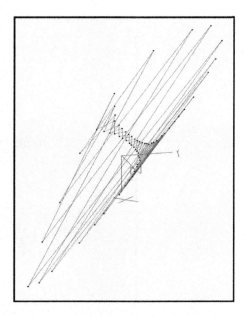

Fig. 4.10 Attractor in the synchronization mode. Control parameters are $k_{xy} = 0.5$, $k_{yx} = 0.4$, k_{yz} $= 0.3$, $k_{zy} = 0.3$, $k_{out} = 0.4$, $p = 0.008$, $q = 0.005$, $r = 0.0057$, $x_{in} = 38$

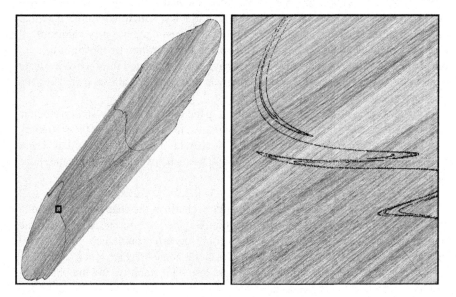

Fig. 4.11 Strange attractor occurring at values of parameters: $k_{xy} = 0.5$, $k_{yx} = 0.1$, $k_{yz} = 0.8$, k_{zy} $= 0.2$, $k_{out} = 0.4$, $p = 0.05$, $q = 0.05$, $r = 0.045$, $x_{in} = 11.6$. On the right is an enlarged fragment, marked with a square in the figure on the left

Fig. 4.12 Strange attractor; parameters are $k_{xy} = 0.1, k_{yx} = 0.1, k_{yz} = 0.1, k_{zy} = 0.1, k_{out} = 0.2, p = 1, q = 1, r = 1, x_{in} = 1.546$

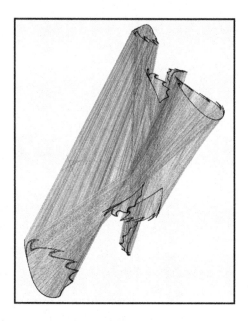

The common iteration procedure was implemented in the algorithm described above. In this procedure, values of dynamic variables, which determine potential energy of the boundaries, change synchronously when there are new values of all three variables calculated on their basis. In a modified algorithm, we took into account the change of values of the variables that have been calculated in order to evaluate values of other variables in the process of given iteration. This allows increasing the accuracy of calculations.

As in the previous algorithm, for small x_{in}, in the system, there is a stationary state defined by the same equation. When x_{in} increases, the system leaves the stationary state. In such a case, a bifurcation similar to the Hopf bifurcation occurs often. There is also a quasiperiodic regime with two aliquant frequencies and attractor in the shape of an annulus (Fig. 4.13).

For the values of parameters $k_{xy} = 0.5, k_{yx} = 0.2, k_{yz} = 0.2, k_{zy} = 0.4, k_{out} = 0.5, p = 1, q = 1, r = 1$, we observed an interesting scenario of the chaos development in the system: From the stationary state after doubling period bifurcation, the periodic regime with an attractor consisting of two points occurs, from which, in its turn, a quasiperiodic regime and an attractor of two annuli come up (Fig. 4.14).

When x_{in} increases, the annuli deform and lose their stability, and the phase trajectory is attracted to the stable attractor consisting of one annulus. We observed a multiple number of attractors in the system, since the attractor in a shape of the annulus had occurred before the attractor consisted of two annuli lost its stability; naturally, each attractor had its own region of attraction. When the bifurcation parameter x_{in} increases, the attractor «annulus» loses its stability and the phase trajectory is attracted to the periodic attractor consisting of two points. In Fig. 4.15, it is shown

Fig. 4.13 Attractor of
quasiperiodic regime. Inside
the annulus is a «web» of
phase trajectories coming
from the stationary
state (a point in the center) to
the attractor ($k_{xy} = 0.5$, k_{yx}
$= 0.1$, $k_{yz} = 0.1$, $k_{zy} = 0.4$,
$k_{out} = 0.5$, $p = 0.05$, q
$= 0.02$, $r = 0.01$, $x_{in} = 30$)

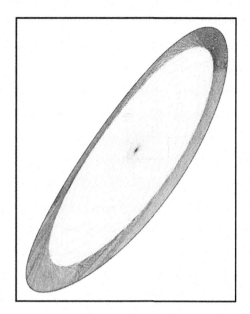

Fig. 4.14 Attractor of quasi
periodic regime with two
aliquant frequencies. It
occurs after periodic attractor
of two points loses its
stability. It is seen how phase
trajectory moves on spirals
from earlier stable attractor
consisting of two points to a
new stable one ($k_{xy} = 0.5$,
$k_{yx} = 0.2$, $k_{yz} = 0.2$, k_{zy}
$= 0.4$, $k_{out} = 0.5$, $p = 1$, q
$= 1$, $r = 1$, $x_{in} = 0.748$)

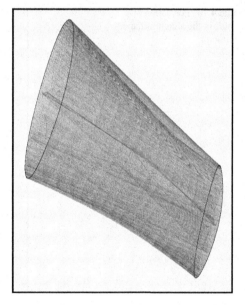

how phase trajectory leaves unstable annulus-shaped attractor (control parameters:
$k_{xy} = 0.5$, $k_{yx} = 0.2$, $k_{yz} = 0.2$, $k_{zy} = 0.4$, $k_{out} = 0.5$, $p = 1$, $q = 1$, $r = 1$, $x_{in} = 0.713$)
and is attracted to the attractor of two points (shown with arrows).

Fig. 4.15 Attractor in
periodic regime consisting of
two points

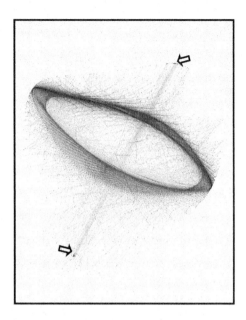

Fig. 4.16 Formation of
folds in the annulus-shaped
attractor

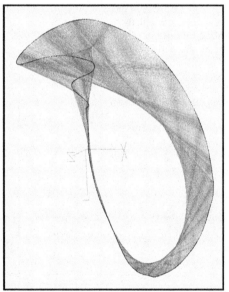

Further increase of x_{in} leads to the occurrence of a chaotic attractor «annulus»
with parameters $k_{xy} = 0.5$, $k_{yx} = 0.2$, $k_{yz} = 0.2$, $k_{zy} = 0.4$, $k_{out} = 0.5$, $p = 1$, $q = 1$, r
$= 1$, $x_{in} = 0.84$. There folds occur on the «annulus», whose number and size increase
with the increasing of x_{in} (Fig. 4.16), and at some moment, a strange attractor comes
up (Fig. 4.17).

Fig. 4.17 The same
attractor when $x_{in} = 0.87$

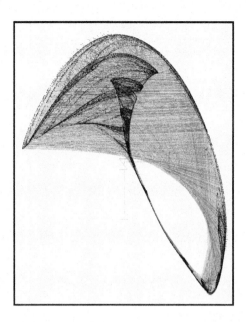

The third modification of the model assumes a more consistent consideration of the energy exchange between the system of internal boundaries and material, and takes into account the possibility of withdrawing energy to internal boundaries, which are not part of the system, from both «extreme» scale levels.

This possibility was modeled by adding to the system one more transition coefficient k_{out1}

$$\Phi_3(x, y, z) = \begin{cases} x_{n+1} = x_n - (k_{xy} + k_{out1})px_n^2 + k_{yx}qy_n^2 + x_{in} \\ y_{n+1} = y_n + k_{xy}px_n^2 - (k_{yx} + k_{yz})qy_n^2 + k_{zy}rz_n^2 \\ z_{n+1} = z_n + k_{yz}qy_n^2 - (k_{zy} + k_{out})rz_n^2 \end{cases} \quad (4.22)$$

$$x_{st} = \sqrt{\frac{x_{in}}{\left[k_{xy} + k_{out1}\left(\frac{k_{yx}}{k_{yz}}\left(\frac{k_{zy}}{k_{out}} + 1\right) + 1\right)\right]p}\left(\frac{k_{yx}}{k_{yz}}\left(\frac{k_{zy}}{k_{out}} + 1\right) + 1\right)};$$

$$y_{st} = \sqrt{\frac{x_{in} - k_{out1}p \cdot x_{st}^2}{k_{yz}q}\left(\frac{k_{zy}}{k_{out}} + 1\right)};$$

$$z_{st} = \sqrt{\frac{x_{in} - k_{out1}p \cdot x_{st}^2}{k_{out}r}} \quad (4.23)$$

The system of equations has an analytic stationary solution:

Fig. 4.18 Strange attractor
of alternation regime (k_{xy}
$= 0.5, k_{yx} = 0.4, k_{yz} = 0.3,$
$k_{zy} = 0.3, k_{out\_1} = 0.7, k_{out}$
$= 0.4, p = 1, q = 1, r = 1,$
$x_{in} = 1.8$)

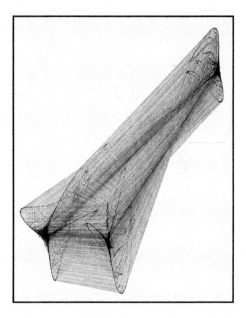

Fig. 4.19 Strange attractor
($k_{xy} = 0.5, k_{yx} = 0.4, k_{yz}$
$= 0.3, k_{zy} = 0.3, k_{out\_1}$
$= 0.65, k_{out} = 0.4, p = 1, q$
$= 1, r = 1, x_{in} = 1.8365$)

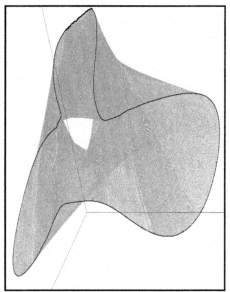

As in the previous cases, we observed two scenarios of transition to chaos. We
observed multiple attractors and alternation of regimes. In Figs. 4.18 and 4.19, we
depict attractors that occurred in the system as an illustration.

4.4 Conclusion

The relations for the calculation of force fields can be applied to the entire range of network sizes, in which there is a self-similarity of internal boundaries. The limits of mesoscopic asymptotics can be determined by the method proposed, for example, in [46, 47].

There are no limitations in the possibility of generalizing the model to internal boundaries located in the body volume: The elements of the Sierpinski square in the «wire» model can be replaced by the edges of cubes of the Menger sponge and its varieties.

As is known, strange attractors can be interpreted as stochastic self-oscillations supported in a dynamic system by an external source. Emerging in the phase space of energies, in which the systems of equations described above are studied, attractors can be interpreted as Lissajous figures of quasi-oscillatory processes. This opens the possibility of a visual determination of the nature and features of the different scale inhomogeneities' interaction.

The coefficients in the system of equations characterize the features of the material structure. As a base for the determination of their numerical values in the first approximation could be the most general assumptions, based on the analysis of the physical situation. The definition of these values for specific composite materials is seen in the way of using knowledge about the structure and properties of force fields of internal boundaries.

References

1. Haken, H.: Synergetic. Introduction and Advanced Topics. Springer, Berlin (2004)
2. Samarskiy, A.A., Zmitrenko, N.V., Kurdyumov, S.P., Mikhailov, A.P.: Teplovyye struktury i fundamental'naya dlina v srede s nelineynoy teploprovodnost'yu i ob"yemnymi istochnikami tepla (Thermal structures and fundamental length in a medium with nonlinear thermal conductivity and volume heat sources). Doklady akademii nauk SSSR (Reports of the Academy of Sciences of the USSR). **227**(2), 321–324 (1976)
3. Herega, A., Volkov, V.: Current status of the inner boundaries: short review the last 30 years. Int. J. Compo. Mater. **4**(5A), 10–15 (2014)
4. Herega, A., Sukhanov, V., Vyrovoy, V.: The model of the long-range effect in solids: evolution of structure clusters of interior boundaries, and their statistical descriptors. AIP Conf. Proc. **1909**, 020069 (2017)
5. Solomatov, V., Bobryshev, A., Proshin, A.: Klastery v strukture i tekhnologii kompozitsionnykh stroitel'nykh materialov (Clusters in the structure and technology of composite building materials). Izvestiya vuzov. Stroitel'stvo (Proceedings of the universities. Construction) **4**, 56–61 (1983)
6. Herega, A.N.: Physical aspects of self-organization processes in composites. 2. The structure and interaction of inner boundaries. Nanomechanics Sci. Technol. **4**(2), 133–143 (2013)
7. Uyomov, A.I.: Sistemy i sistemnyye parametry (Systems and system parameters). In: Problemy formal'nogo analiza sistem (The problems of formal analysis of systems). Vysshaya shkola, Moscow (1968)
8. Von Bertalanffy, L.: General system theory. A critical review. In: General Systems, VII (1962)

9. Herega, A., Vyrovoy, V.: Inter'yernyye granitsy kompozitov: polimasshtabnost' struktury i svoystva silovykh poley (Interior boundaries of composites: the multiscale structure and the properties of force fields). In: Abstracts of the Proceeding of the 4th All-Russian Symposium of the Mechanics of Composite Materials and Structures, Institute of Applied Mechanics RAS, Moscow, 4–6 December 2012
10. Holland, J.: Complexity: a very short introductions. Oxford University Press, Oxford (2014)
11. Panin,V.Y., Grinyayev, Y.V., Danilov, V.I.: Strukturnyye urovni plasticheskoy deformatsii i razrusheniya (Structural levels of plastic deformation and fracture). Nauka, Novosibirsk (1990)
12. Olemskoi, A.I., Sklyar, I.A.: Evolution of the defect structure of a solid during plastic deformation. Sov. Phys. Usp. **35**(6), 455–480 (1992). https://doi.org/10.1070/pu1992v035n06abeh002241
13. Herega, A., Ostapkevich, M.: Computer simulation mesostructure of cluster systems. AIP Conf. Proc. **1623**, 209–212 (2014)
14. Herega, A., et al.: Percolation model of composites: fraction clusters and internal boundaries. Int. J. Compo. Mater. **2**, 142–146 (2012)
15. Psakh'ye, S.G., Zol'nikov, K.P., Kryzhevich, D.S.: Calculation of diffusion properties of grain boundaries in nanocrystalline copper. Phys. Mesomech. **1**, 25–28 (2008)
16. Morozov, Yu., Simbukhov, I., Dyakonov, D.: Study of microstructure and properties of ultrahigh-strength pipe steel of strength category x120 prepared under laboratory conditions. Metallurgist **56**, 510–518 (2012)
17. Turchenko, V. et al: Structural features, magnetic and resistive properties of nanoparticle perovskites, prepared by sol gel method. In: Proceedings of IV International Conference of Functional Nanomaterials and High-purity Substances. Suzdal, 1–5 October 2012
18. Salamon, M., Jaime, M.: The physics of manganites: structure and transport. Rev. Mod. Phys. **73**, 583–628 (2001)
19. Fedotov, A., Mazanik, A., Ulyashin, A.: Electrical activity of grain boundaries in silicon bicrystals and its modification by hydrogen plasma treatment. Sol. Energy Mater. Sol. Cells **72**, 589–595 (2002)
20. Fedotov, A., Mazanik, A., Katz, E., et al.: Electrical activity of tilt and twist grain boundaries in silicon. Solid State Phenom. **67–68**, 15–20 (1999)
21. Reda, Chellali M., Balogh, Z., Schmitz, G.: Nano-analysis of grain boundary and triple junction transport in nanocrystalline Ni/Cu. Ultramicroscopy **132**, 164–170 (2013)
22. Zhu, M.W., Wang, Z.J., Chen, Y.N., et al.: Effect of grain boundary on electrical properties of polycrystalline lanthanum nickel oxide thin films. Appl. Phys. A **112**, 1011–1018 (2013)
23. Hirose, M., Tsunemi, E., Kobayashi, K., Yamada, H.: Influence of grain boundary on electrical properties of organic crystalline grains investigated by dual-probe atomic force microscopy. Appl. Phys. Lett. **103**, 109–112 (2013)
24. Mileiko, S.: Composites and Nanostructures **1**, 6 (2009)
25. Panin, V., Fomin, V., Titov, V.: Fizicheskiye printsipy mezomekhaniki poverkhnostnykh sloyev i vnutrennikh granits razdela v deformiruyemom tverdom tele (Physical principles of mesomechanics of surface layers and internal interfaces in a deformable solid). Fizicheskaya mezomekhanika (Phys. Mesomech.) **6**, 5–14 (2003)
26. Cherepanov, O., Pribytkov, G.: Chislennoye issledovaniye uprugoplasticheskikh deformatsiy metallokeramiki pri zakalke (Numerical study of elastoplastic strains of cermet during quenching). Fizicheskaya mezomekhanika (Phys. Mesomech.) **3**, 33–43 (2000)
27. Psakh'ye, S.G., Uvarov, T.Y., Zol'nikov, K.P.: O novom mekhanizme generatsii defektov na granitsakh razdela. Molekulyarno dinamicheskoye modelirovaniye (On a new mechanism for the generation of defects at interfaces. Molecular dynamic modeling). Fizicheskaya mezomekhanika (Phys. Mesomech.) **3**, 21–23 (2000)
28. Psakh'ye, S.G., Zol'nikov, K.P., Dmitriev, A.I., Smolin, AYu., Shil'ko, Y.V.: Dynamic vortex defects in deformed material. Phys. Mesomech. **17**, 15–22 (2004)
29. Makarov, P.V.: Podkhod fizicheskoy mezomekhaniki k modelirovaniyu protsessov deformatsii i razrusheniya (The approach of physical mesomechanics to the modeling of deformation and fracture processes). Fizicheskaya mezomekhanika (Phys. Mesomech.) **1**, 61–81 (1998)

30. Panin, V.Y., Moiseyenko, D.D., Yelsukova, T.F.: Mnogourovnevaya model' deformiruyemogo polikristalla. Problema Kholla-Petcha (Multilevel model of a deformable polycrystal. The Hall-Petch problem). Fizicheskaya mezomekhanika (Phys. Mesomech.) **16**, 15–28 (2013)
31. Psakh'ye, S.G., Astafurov, S.V., Shil'ko, E.V.: Effect of elastic characteristics of the surface layer on the deformation properties of materials with an interface-controllable structure. Tech. Phys. **52**, 1523–1526 (2007)
32. Yanovsky, Yu., Obraztsov, I.: Some aspects of computer modeling of advanced polymer composite materials structure and micromechanical properties. Phys. Mesomech. **1**, 129–135 (1998)
33. Kozlov, G., Yanovskii, Yu., Zaikov, G.: Structure and Properties of Particulate-Filled Polymer Composites. Nova Science Publishers, New York (2010)
34. Kozlov, G.V.: Structure and properties of particulate-filled polymer nanocomposites. Phys. Usp. **58**, 33–60 (2015). https://doi.org/10.3367/ufne.0185.201501c.0035
35. Lifshitz, I.M., Kosevich, A.M.: The dynamics of a crystal lattice with defects. Kept. Progr. Phys. **29**(1), 217–254 (1966)
36. Bozhokin, S.V., Parshin, D.A.: Fraktaly i mul'tifraktaly (Fractals and Multifractals). R&H Dynamics, Moscow-Izhevsk (2001)
37. Mandelbrot, B.: The Fractal Geometry of Nature. W. H. Freeman and Co., San Francisco (1982)
38. Herega, A., Drik, N.: Ugol'nikov A.: Hybrid ramified Sierpinski carpet: percolation transition, critical exponents, and force field. Physics-Uspekhi **55**(5), 519–521 (2012). https://doi.org/10.3367/UFNe.0182.201205f.0555
39. Mandelshtam, L.I.: Lektsii po teorii kolebaniy (Lectures on the Theory of Oscillations). Nauka Press, Moscow (1972)
40. Broek, D.: Elementary Engineering Fracture Mechanics. Nijhoff, Hague (1982)
41. Andronov, A.A., Vitt, A.A., Khaykin, S.E.: Teoriya kolebaniy (Theory of oscillations). Nauka, Moscow (1981)
42. Bekker, M., Herega, A., Lozovskiy, T.: Strange attractors and chaotic behavior of a mathematical model for a centrifugal filter with feedback. Adv. Dyn. Syst. Appl. **4**, 179–194 (2009)
43. Feigenbaum, M.J.: Universal behavior in nonlinear systems. Los Alamos Sci. **1**, 4–27 (1980)
44. Bergé, P., Pomeau, Y., Vidal, C.: L'ordre dans le chaos. Hermann, Paris (1997)
45. Lichtenberg, A.J., Lieberman, M.A.: Regular and Stochastic Motion. Springer, Berlin, Heidelberg (1992)
46. Sokolov, I.M.: Dimensions and other critical indictors in the percolation theory. Sov. Phys. Usp. **29**, 924–945 (1986). https://doi.org/10.1070/PU1986v029n10ABEH003526
47. Aslanov, A.M., Bekker, M.B., Vyrovoj, V.N., Herega, A.N.: Imitation model of synergetic processes in dynamic disperse systems: Ξ Criterion. Tech. Phys. **55**, 147–150 (2010)

Chapter 5
Small Clusters: Genesis, Structure, and Properties

Abstract The fourth chapter of the monograph describes a simulation computer model of coagulation of a solid phase of a circulating two-phase gas stream under conditions of centrifugal force. The model makes it possible to study the structure and properties of forming clusters, the effect of the dispersed phase and the effect of the structure in which the motion takes place, the role of particle interaction, and the influence of chaotic flow parameters. In the monograph types of model, clusters are described, their structure and habitus are studied; the possibility of a priori determination of the type of forming clusters as well as the type of trajectories and trends of flow development on the basis of the multidimensional Ξ-criterion are shown. The model shows that obtained clusters are multifractals; the conditions of their stability are investigated, and also the theorem on the field interaction of mesoscopic clusters, which allows us to describe the interactions of arbitrary shape bodies at finite distances, is proved. As a development of the model, a method for reconstructing flow properties from geometric characteristics of solid phase particles imprints on walls of the channel structure is proposed. For this purpose, a method that substantially simplifies the analysis of such a problem and allows us to move from the consideration of a dynamic flow to the analysis of a stationary spot to reduce the dimensionality of the object under study is proposed. This is possible due to the presence (under certain conditions) of the correlation between the characteristics of the spots and the flow, which, in particular, is expressed by the dependence between the fractality of the first and the persistence of the second.

What's more, a methodology for the quantitative evaluation of the ordering degree of such an image has been developed; methods for determining the degree of randomness and the level of persistence of the flow are described; a question of the influence of these factors on the structure of model clusters is investigated. In addition, a method for determining the boundaries of the intermediate asymptotic of the cluster structure is proposed, and a technique for describing the interaction of arbitrary shape bodies at finite distances is developed.

The chapter describes a simulation computer model of coagulation of a solid phase of a circulating two-phase gas stream under conditions of centrifugal force. Types of model clusters are described, and their structure and habitus are studied; the

A. Herega, *The Selected Models of the Mesostructure of Composites*, SpringerBriefs in Physics, https://doi.org/10.1007/978-3-319-89704-2_5

53

possibility of a priori determination of the type of forming clusters as well as the type of trajectories and trends of flow development on the basis of the multidimensional Ξ-criterion is shown. The dependence of the clusters' fractal dimension on the parameter determined by structural characteristics of the channel is obtained.

As a development of the model, a method for reconstructing flow properties from geometric characteristics of solid phase particles imprints on walls of the channel structure is proposed; a methodology for the quantitative evaluation of the ordering degree of such an image has been developed; methods for determining the degree of randomness and the level of persistence of the flow are described; a question of the influence of these factors on the structure of model clusters is investigated. In addition, a method for determining the boundaries of the intermediate asymptotic of the cluster structure is proposed, and a technique for describing the interaction of arbitrary shape bodies at finite distances is developed.

5.1 Simulation of Coagulation in a Curvilinear Two-Phase Flow. Types of Clusters. Ξ-Criterion

Aerosol is a colloidal suspension of particles dispersed in air or gas. The continuous and spontaneous coagulation of aerosols is one of their most interesting properties. When colliding, particles of any material coalesce or stick together, making the aerosol «coarser» [1]. In disperse systems, coagulation is known to proceed under the action of van der Waals and gravitational forces [1–5]. Van der Waals forces are short-range electrostatic attractive forces between uncharged molecules, arising from the interaction of electric dipole moments. van der Waals forces, as is well known, can have three components: dipole–dipole (V. Keesom forces), induction (P. Debye's forces), and dispersive (F. London's forces) [3–5]. Keesom forces provide an orientational interaction between two permanent dipoles. (Dipole or dipole moment is the mathematical product of the separation of the ends of a dipole and the magnitude of the charges.); they manifest in the fact that polar molecules tend to orient during the approach to each other oppositely. Debye force results from the interaction between the permanent dipoles of polar molecules and the dipoles they may induce in non-polar molecules. London force is a temporary attractive force that results when the electrons in two adjacent atoms occupy positions that make the atoms form temporary dipoles; this force is the weakest intermolecular force.

The coagulation process is also influenced by the concentration, size, and shape of particles; polydispersity of the composition (dispersion is a characteristic of the particle size in disperse systems (colloidal solutions, sols, etc.); measure of dispersion is the ratio of the total surface of all particles to their total volume or mass. Polydispersity is determined by the function of particle distribution over size or mass); rate of particle mixing in the flow; and interaction with the wall of a construction, etc. Experimental investigation of clusters forming in a gas is therefore a challenge:

Researchers often fail in achieving a desired range of structure parameters and lack reliable methods of diagnostics thereof [6–9].

Analytical description of such a multifactor process is also not an easy task. For an instance, even if in some approximation the kinetic equation describing coagulation of particles in a two-phase flow (two-phase gas flow is a stream containing solid particles, widely used in various technological processes and in scientific research) has a relatively simple form

$$\frac{dn}{dt} = -\left(Kn^2 + \beta n\right), \tag{5.1}$$

where K is a coagulation constant associated with the known van der Waals-Hamaker constant, β is a factor describing the interaction of solid stream particles with walls of the construction [4], then it is impossible to determine these constants for chaotic curvilinear flows by analytical methods.

In this regard, high hopes are placed on theoretical studies, including simulation and statistical modeling. In [10], it is noted that «apparently, the most reliable predictions of such systems properties are given only by computer calculations, which carried out, in particular, by the Monte Carlo method». In such a situation, simulation as a modeling process is a useful method for solving the problem.

Aggregate (in physical phenomena) is the accumulation of small particles that formed a larger one. Fractal aggregate is an object that is formed during the association (agglutination) of solid aerosols in the case of their diffusive motion in a gas or liquid. Modern findings on fractal aggregates, based largely on theoretical models, do not always correspond to experiments [6, 11–13]. One of the reasons is that in analytical and numerical studies the nature of an aggregation process, a motion type of clusters and particles, a probability of coagulation in interaction is fixed in advance [6]. What's more, the description of the interaction with the construction is maximally simplified or ignored.

A proposed simulation statistical model was created to study processes of a cluster formation under the action of centrifugal forces. In the model, which in many ways generalizes researches [14–20], the above described typical for this type models drawbacks are successively eliminated. It allows us to investigate the structure and properties of model clusters, to study the influence of characteristics of the dispersed phase and the chaotic flow, to determine the role of the interaction of particles and clusters with each other and with the walls of the device.

The model is an algorithm of the motion and interaction of solid particles in a flow. In a computer experiment, a chaotic motion of particles in a circulating flow in a curvilinear channel of a square section is modeled. Each act of moving the particle is associated with a vector having a random, transferable, and also a component determined by centrifugal force. Each model experiment involved up to 5000 particles. It is assumed that particle sizes are distributed according to the normal law, and their adhesion occurs when approaching a distance equal to some minimum. A value of the interaction in the model is determined by masses of particles and clusters and by the distance between them. The destruction of clusters is caused by a value of mechanical stresses in them. In the interaction with the construction particles and clusters behave like solid spheres.

Experiments are terminated when the number of particles in the coagulant reaches 100. In each of the regimes, the experiment was repeated 50 times, so that the results were accurate to within 3–5%.

Analysis of results showed the presence of a competition in the action of listed factors. It turned out that both an increase in the flow velocity and an increase in centrifugal force (at a constant velocity) lead to an increase in the fractal dimension of the aggregate, while an increase in the flow width lead to its decrease [14].

In view of this, it is convenient to consider the coagulant fractal dimension as a function of the ratio between the mean kinetic energies of the transverse (azimuthally) and radial motions of the particles. This is equivalent to the ratio of the radius of curvature of the channel to the flow width,

$$\alpha = \frac{E_{k.t}}{E_{k.r}} = \frac{mV^2}{(mV^2/r_{cur})d_f} = \frac{r_{cur}}{d_f}, \tag{5.2}$$

where $E_{k.t}$, $E_{k.r}$ are the average values of the kinetic energy of the transverse and radial component of the chaotic motion of coagulants, m is the mass of the particle, V is its transverse velocity, r_{cur} is the radius of the channel curvature, and d_f is the width of the air flow.

When the parameter α is verified from values of the coagulants fractal dimension obtained in simulating experiments where r_{cur} and d_f were varied in such a way that α remained constant, it turned out that the spread of corresponding values of the fractal dimension does not exceed 5%. The coagulant fractal dimension D as a function of parameter α in the model suggested can be expressed as $D(\alpha) = 0.73(1 + \alpha^{0.18})$.

Thus, it was shown the principal possibility of obtaining a monotonic dependence of the coagulant fractal dimension on the parameter describing constructional characteristics of the channel.

Model experiments also showed that the structure and density of clusters with fixed interaction parameters depend on the type of particle distribution by sizes and the degree of flow randomness.

There are two types of coagulants that are formed in the model: quasisymmetric, which are usually rather compact, and asymmetric with a lower density (Figs. 5.1 and 5.2). Model experiments showed that the predominant type of clusters in the flow can be determined a priori on the basis of the proposed Ξ-criterion [15].

Let the number Ξ be a ratio of the mean values of transferable V_{trans} and random V_{chaot} components of the flow velocity

$$\Xi = V_{trans}/V_{chaot}. \tag{5.3}$$

Model experimental data showed that for $\Xi < 0.425$ the probability of asymmetric model clusters occurrence tends to zero; for $\Xi > 0.575$, almost all forming clusters are asymmetric.

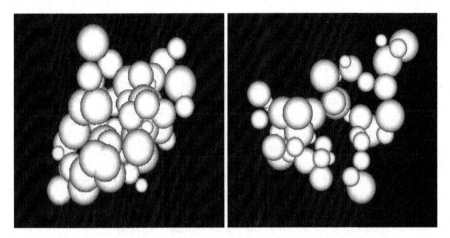

Fig. 5.1 Asymmetric model clusters

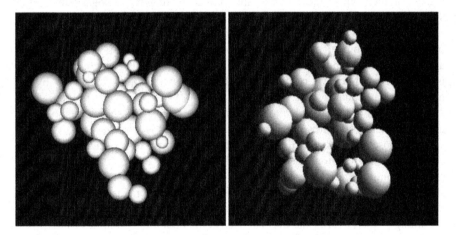

Fig. 5.2 Quasisymmetric clusters

This has an obvious qualitative explanation: If the transferable velocity is relatively low, particles can penetrate very deep into the cavities of the cluster. Therefore, in this case, only the quantitative component of the result is of interest—the value of the number Ξ as the predominance of one of cluster types when the velocity component ratio takes an extreme value is quite predictable.

The habit of model clusters gives grounds for assuming the presence in their structure of a statistical self-similarity, which, as we know, implies that the random values characterizing the cluster particles arrangement have same statistical distributions and equal mathematical expectations and variances [21]. The model confirmed the assumption that obtained clusters are multifractals: heterogeneous fractal objects that require an infinite set of measures for an adequate description [22]. The need for such a set is due to the fact that the structure of multifractals has a spectrum of scaling, i.e., is determined not by one but by a set of scale factors. In this case, the object is usually characterized by the spectrum of generalized Renyi dimensions (see Appendix C). It was shown in [22] that Renyi dimensions of the multifractal form a series $D_0 \geq D_1 \geq D_2 \geq D_3 \ldots$, and equality takes place if the object is the fractal: In this case, all dimensions are equal to the fractal dimension D_0.

To calculate the fractal dimension, the model uses a volume analogue of a software programmable reticulation (it is a grid of squares on a transparent plate, paper, glass; as a rule, it is used to determine the areas of sites on geographic maps, and sometimes, for scientific purposes, for example, to determine different dimensions) [22]. Calculation of the leading dimensions of the spectrum, which requires determination of the probability that an object under study is in a cell of the reticulation, was carried out by the Monte Carlo method.

The method of dimensions calculating is standard: Graphs of the dependence of the number of volume reticulation cells, which contain elements of the studied cluster, on the value of the cell edge are constructed in a double logarithmic scale. The graphs have a characteristic appearance and consist of two almost horizontal sections and one that is at an angle. It is known that it is it that corresponds to the intermediate asymptotics [22]; the model calculates the tangent of a slope angle of an inclined section approximating straight line (see Fig. 5.3b), which, up to a sign, is equal to the dimension of the object [22]. Approximation is carried out by the method of least squares [23].

An alternative way to calculate cluster dimensions is the sphere method. It is also implemented in software. The calculation process is iterative: It starts with a sphere, whose radius is several characteristic particle sizes that form the cluster, and each subsequent iteration increases the sphere radius by a fixed value. The process continues until the next sphere will not «cover» the extreme cluster particle. At each step, the cluster volume contained in the sphere of a current radius is calculated. The numerical value of the dimension is determined, as in the case of the reticulation, according to the graph of the dependence (in the double logarithmic scale) of this volume on the sphere radius.

The application of the sphere method to calculate the three-dimensional clusters dimensions gave also a serendipity result (it is a term denoting ability, when drawing conclusions from observations, to find something that was not sought intentionally). It turned out that the method has limitations: It is correct to carry out calculations only for quasisymmetric clusters. In this case, obtained values of the dimension differ from results obtained by the reticulation method by $\pm(3 \div 7)\%$. When calculating dimensions of asymmetric clusters, the sphere method shows results for which physical interpretation is impossible.

If the cluster is not a fractal (in the model, this is possible for two reasons: The cluster is small, and self-similarity does not show up, or the cluster has a strongly marked asymmetry), this is seen from the arrangement of points in the intermediate asymptotics of the graph constructed to determine the fractal dimension. In the case of the fractal (multifractal) cluster, experimental points are easily approximated by the straight line; otherwise, they spread in the form of a «cloud» (Fig. 5.3). Boundaries

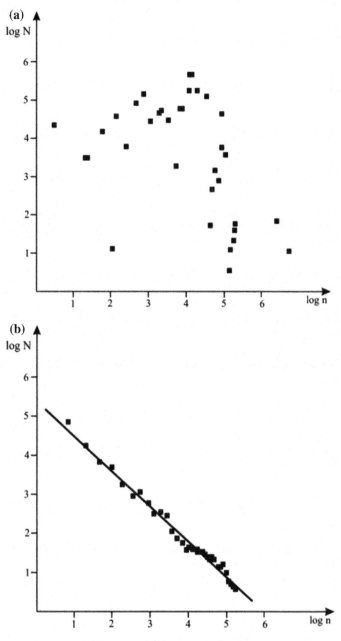

Fig. 5.3 View of the region of intermediate asymptotics for nonfractal (**a**) and multifractal (**b**) model clusters

Table 5.1 Renyi dimensions for one realization of model clusters

Dimension	Quasisymmetric clusters	Asymmetric clusters
D_{max}	$2.98 \pm 0.05 \; \varepsilon = 1.7\%$	$2.75 \pm 0.06 \; \varepsilon = 2.2\%$
D_{fract}	$2.90 \pm 0.03 \; \varepsilon = 1.0\%$	$2.67 \pm 0.08 \; \varepsilon = 3.0\%$
D_{inf}	$2.73 \pm 0.04 \; \varepsilon = 1.4\%$	$2.54 \pm 0.04 \; \varepsilon = 1.6\%$
D_{corr}	$2.67 \pm 0.02 \; \varepsilon = 0.7\%$	$2.47 \pm 0.09 \; \varepsilon = 3.6\%$
D_{third}	$2.59 \pm 0.03 \; \varepsilon = 1.2\%$	$2.34 \pm 0.02 \; \varepsilon = 0.9\%$
D_{fourth}	$2.50 \pm 0.05 \; \varepsilon = 2.0\%$	$2.29 \pm 0.04 \; \varepsilon = 1.7\%$
D_{min}	$2.31 \pm 0.02 \; \varepsilon = 0.9\%$	$2.11 \pm 0.05 \; \varepsilon = 2.4\%$

of the intermediate asymptotics for different dimensions of the spectrum are not the same: The information D_{inf} and maximal D_{max} dimensions (see Appendix C) turned out to be the most sensitive to the loss of self-similarity by the cluster. Therefore, the other five dimensions were calculated between them.

Table 5.1 presents results of statistical processing of obtained values of Renyi dimensions of model clusters containing not less than one hundred particles. They confirm the assumption of the clusters multifractal nature [22].

5.2 Stability of Mesoscopic Clusters. The Theorem on the Field Interaction of Multifractals

The model assumes that the cluster is destroyed by contact interaction with a particle, with another cluster, or with a channel wall only if the mechanical stresses in the cluster exceed a certain limiting value, which is used as one of the model control parameters. Thus, the problem of calculating the fields of mechanical stresses in the fractal cluster arises. The solution of this problem allows describing interactions of arbitrary shape bodies at finite distances.

Pioneer works on the description of the regular geometric shape bodies' interaction at finite distances belong to Zhukovsky [24]. The algorithm for constructing a force field proposed in the model is the modernization of the Zhukovsky method for a case when an object is a fractal cluster. The algorithm is based on the analogue of the Zhukovsky's theorem [24] on the attraction of a point by a material plate of finite sizes.

Theorem *A volume multifractal object of finite sizes attracts a material point of unit mass with force $F = a \cdot \sum \rho_i \omega_i$, where a is a constant that depends on the interaction type, ρ_i is a local bulk density, and ω_i is a solid angle, under which a corresponding fragment of the multifractal from the attracted point is visible.*

The proof of the theorem is analogous to the one given by Zhukovsky [24].

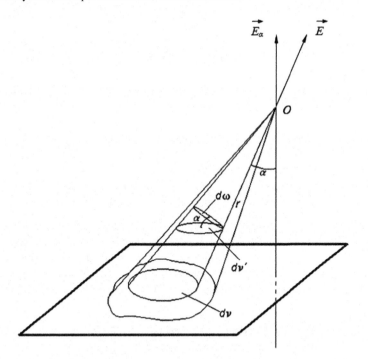

Fig. 5.4 To the proof of the theorem on the interaction of the multifractal with the material point

Let there be an infinitely thin fractal plate of an arbitrary shape and finite sizes. And
let there be an arbitrary point O of unit mass. Let us determine a normal component
of a stress vector, which determines interaction force (Fig. 5.4).

For this reason, let us fall a perpendicular from the point O on the plate and match
z-axis with it. Then let us select an element of volume dv and construct a small cone
by joining the contour of dv with the point O. Force of attraction by the element dv
of the point O, in fact, the field stress, is

$$dE(r) = a \cdot \rho(r) \cdot dv/r^2 \tag{5.4}$$

where a is the coefficient of interaction, $\rho(r)$ is the local density, dv is the volume
element, and r is a distance from the point O to the element dv.

Now let us transform the expression (5.4). We construct in the point O a sphere of
unit radius that cuts out an area element dw from the cone mentioned above (Fig. 5.4).
If we intersect the cone by a plane parallel to the plate surface and tangent to dw, we
obtain the area element dv'. Since areas of cone parallel sections are referred to as
squares of distances to the vertex, we have

$$dv'/dv = 1/r^2, \text{ or } dv = r^2 \cdot dv', \tag{5.5}$$

here dv is a contour of the volume element.

But $dw = dv'\cos\alpha$, where α is an angle between the normal line and the direction to the fragment under consideration, in other words, between perpendiculars to the contours dv' and dw, consequently, the element contour $dv = r^2 \cdot dw/\cos\alpha$. Then, for the normal component of the stress generated by the element dv and influencing on an arbitrary point of unit mass, we obtain

$$\mathrm{d}\,\mathrm{En} = a \cdot \rho(r) \cdot \cos\alpha \cdot \mathrm{d}v/r^2 = a \cdot \rho(r) \cdot \mathrm{d}w, \qquad (5.6)$$

where dw is a solid angle, under which the contour of the volume object is visible. Quod erat demonstrandum.

The algorithm for constructing force fields based on the theorem is simple:

- Around the object a straight parallelepiped is constructed, divided into cubic cells, the size of which is determined by the necessary accuracy of results;
- From the point of space in which the stress is determined, a solid angle, under which the fragment of the multifractal is visible, is constructed. For this angle, the value of dE is calculated. Then, summation over the parallelepiped cells is performed.

Intracluster Interaction. The evolution of fractal clusters in a chaotic flow is known to be a combination of division and growth. One of the possible causes of destruction is increased fragility. Consider the dependence of this property on the cluster shape, and hence on the conditions for its formation.

Let us assume that interaction forces between particles of the cluster are proportional to r^{-2}. It is clear that in this case the nearest neighbors of the particle will contribute an exponentially large contribution to the stress value.

These divergent terms are eliminated in the model by a simple rule: The particles located at a distance of 2–3 average particles radii from the point, in which the field is computed, are excluded from computation of stress in it. In the scope of the model, this brings about the fact that the experimental dependence $E(r)$ is approximated by a linear function inside a cluster and by a function $1/r^2$ at a distance of 5–10 radii from it (Fig. 5.5).

It is this character of the dependence $E(r)$ that allows introducing the notion of an effective radius of the cluster, which is the radius of the sphere that can substitute it when describing power interactions. If one compares it to the classical result, which is the field of a solid sphere, it can be seen that the effective radius is determined by the coordinate of the graph kink.

Model experiments have shown that the value of mechanical stresses in quasisymmetric clusters is, on average, 3–5 times less than in asymmetric clusters.

Randomness of Flow and Properties of Clusters. *The Interaction of the Flow with the Construction*

Interaction of a two-phase flow with the wall of the construction is not only an important part of the motion but also a factor that aids in reconstructing by solid

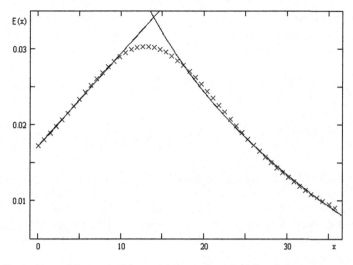

Fig. 5.5 Dependence of the stress of the model cluster force field (in relative units) on the distance

phase clusters' imprints some of flow characteristics. In fact, this approach realizes an attempt to reconstruct a process from its «shadows»,[1] in so doing, substantially simplifying the problem and allowing, firstly, replacing the dynamic flow by a stationary spot, and, secondly, reducing the dimension of the studied object. Such an approach is possible because the characteristics of the spots and flow correlate (under certain conditions); that is, the fractality of the former is related to the persistence of the latter.

A calculation algorithm used in the model goes back to the method of normalized range R/S of a random series of events $f(t)$ for which the Hurst empirical law is fulfilled

$$R/S \sim (\tau/2)^H, \tag{5.7}$$

Where

$$R(\tau) = \max_{1 \le t \le \tau} h(t, \tau) - \min_{1 \le t \le \tau} h(t, \tau), \tag{5.8}$$

$$h(t, \tau) = \sum_{u=1}^{t} \{f(u) - f_\tau\}, \tag{5.9}$$

---

[1]In a dialogue «The Republic» Plato writes: «People seem to be in an underground dwelling, like a cave ... They are turned with their backs to the light. Such prisoners would completely and entirely accept shadows of objects carried by them as the truth» [25].

S is the standard deviation, τ is the length of the time interval we are interested in, and H is the Hurst index [22, 26]. The value of H in the model is reconstructed from fractal dimension of the spot D and fractal dimension of the space d.

$$H = d - D. \tag{5.10}$$

If H is other than 0.5, random quantities of the process correlate: For $H > 0.5$, the process is persistent (persistence is the preservation of the current trend in development, for example, in stochastic processes); for $H < 0.5$, the process is antipersistent [22, 26].

The idea of reconstructing the properties of the flow from stable imprints has various applications. Indeed, in [27], a method for studying a radioactive impurity spot formed on the surface of the Earth is described. The structure of this spot is determined by the properties of turbulent airflow. And in [28], used in studying the processes of turbulent combustion, a method of trace imprints of gas detonation on the inner surface of the pipe, in which the process develops, is described.

To reconstruct the properties of the flow from imprints, it is necessary to gain insight into the interaction of the flow with the construction. It seems obvious that sizes of clusters interacting with the construction walls should be taken into account in simulation of the spot. To do this, one can, e.g., construct a sphere into which a cluster is inscribed and also determine a distance from its center at which a section is passed to find an appropriate surface area of the imprint. This distance could be used as an adjustable parameter of the model. However, model experiments showed that the «pictures» of imprints left by the centers of mass of clusters are most informative, i.e., most sensitive to the flow features.

A peculiarity of this motion is that the flow passes over the boundary surface many times. Accordingly, the spot evolves and the information content of its characteristics and the reconstruction quality of generating flow properties change. This requires determining the conditions under which the reconstruction quality is maximized.

A criterion that a spot is ready for analysis of flow properties is its self-similarity (acquirement of multifractal properties). Computer experiments showed that when the coverage of the construction wall is 25–45%, the «pattern» of imprints is already a multifractal, the Renyi dimension spectrum of which can be found correctly. In this case, the Hurst index determined by formula $H = d - D$ may be both larger and smaller than 0.5 depending on flow properties [15]. The data presented in Table 5.2 refer to the antipersistent process: Even if spots cover a considerable surface area, any distinct trend in the flow evolution is absent and their pattern is structured only slightly [22, 27].

The above method of finding the scale on which spatial bodies possess these properties can also be extended for spots. (Questions of estimation of the ordering degree (structuring) of images are considered in Appendix D).

Based on the results of model experiments, one can see that the number Ξ can be a control parameter that changes the value of the Hurst index. Ξ is the ratio of the mean values of transferable V_{trans} and random V_{chaot} components of the gas flow

Table 5.2 Renyi dimensions for one realization of the multifractal imprint of the solid phase on the wall of the construction

Dimension	Value
D_{max}	1.93 ± 0.05 $\varepsilon = 2.6\%$
D_{fract}	1.90 ± 0.03 $\varepsilon = 1.6\%$
D_{inf}	1.71 ± 0.06 $\varepsilon = 3.5\%$
D_{corr}	1.69 ± 0.09 $\varepsilon = 5.3\%$
D_{third}	1.51 ± 0.09 $\varepsilon = 6.0\%$
D_{fourth}	1.47 ± 0.02 $\varepsilon = 1.4\%$
D_{min}	1.36 ± 0.06 $\varepsilon = 4.4\%$

Table 5.3 Multiaspect Ξ-criterion

Value of the criterion	V_{tran}/V_{chaot}	Clusters	Trajectories	Imprint	Flow
$\Xi \leq 0.425$	~2/5	Quasisymmetric, compact	Asymptotically unstable	Expressed structured	Persistent
$\Xi \geq 0.575$	~3/5	Asymmetric, non-coherent	Asymptotically stable	Almost homogeneous	Antipersistent

velocity and is the constant of the model [15]: at $\Xi < 0.425$, Hurst index $H > 0.5$; at $\Xi > 0.575$, $H < 0.5$.

In the model, the asymptotic Lyapunov stability of trajectories was determined [29]. It is shown that the same values of criterion Ξ that indicate the presence or absence of evolutionary trends in the flow and the types of clusters prevailing in the model flow [15] determine also the signs of Lyapunov exponents. At $\Xi > 0.575$, Lyapunov exponents in the model are determined by the signature <0, −, −> which is known to indicate the presence of a limit cycle in the behavior of the distance between nearby trajectories; at $\Xi < 0.425$, they are determined by the signature <+, 0, −> which suggests the presence of a strange attractor [29]. No distinct tendency was revealed between these values.

Thus, the type of clusters prevailing in a model flow, the presence of evolutionary trends, and the type of flow trajectories behavior can be determined a priori using the Ξ criterion suggested (see Table 5.3).

References

1. Green, H., Lane, W.: Particulate clouds: dusts, smokes and mists. E. & F. Spon Ltd, London (1964)
2. Nakamura, H., Deguchi, N., Takeuchi, H., Watano, S.: Numerical analysis of fluid flow and particle entrainment in a novel tapered rotating fluidized bed. Chem. Eng. Sci. **116**, 725–733 (2014)

3. Sonntag, H., Strenge, K.: Coagulation kinetics and structure formation. Springer, Berlin (1987)
4. Yavorsky, B., Detlaf, A.: Spravochnik po fizike (Handbook of physics). Mir, Moscow (1980)
5. Levich, V.G., Vdovin, Yu.A., Myamlin, V.A.: Kurs teoreticheskoj fiziki (Course of theoretical physics), vol. 2. Nauka, Moscow (1971)
6. Mikhailov, E.F., Vlasenko, S.S.: The generation of fractal structures in gaseous phase. Phys. Usp. **38**, 253–271 (1995). https://doi.org/10.1070/PU1995v038n03ABEH000074
7. Petrov, Yu.I.: Klastery I malye chastitsy (Clusters And Small Particles). Nauka, Moscow (1986)
8. Witten, T., Sander, L.: Diffusion limited aggregation: a kinetic critical phenomenon. Phys. Rev. Lett. **47**, 1400–1485 (1981)
9. Pietronero, L., Tosatti, E. (eds.): Fractals in physics. North-Holland, Amsterdam, New York (1986)
10. Martin, F.F.: Computer modeling and simulation. Wiley, New York (1968)
11. Jullien, R.: Fractal aggregates. Comm. Cond. Mat. Phys. **13**, 177–205 (1987)
12. Smirnov, B.M.: Properties of a fractal aggregate. Sov. Phys. Usp. **32**, 181–182 (1989). https://doi.org/10.1070/PU1989v032n02ABEH002682
13. Smirnov, B.M.: Fractal clusters. Sov. Phys. Usp. **29**, 481–505 (1986). https://doi.org/10.1070/PU1986v029n06ABEH003414
14. Aslanov, A.M., Herega, A.N., Lozovskij, T.L.: Two models of stochastic processes in centrifugal feedback filters. Tech. Phys. **51**(6), 812–813 (2006)
15. Aslanov, A.M., Bekker, M.B., Vyrovoy, V.N., Herega, A.N.: Imitation model of synergetic processes in dynamic disperse systems: Ξ criterion. Tech. Phys. **55**, 147–150 (2010)
16. Bekker, M., Herega, A., Lozovskiy, T.: Strange attractors and chaotic behavior of a mathematical model for a centrifugal filter with feedback. Adv Dyn Syst Appl **4**(2), 179–194 (2009)
17. Herega, A.N.: Physical aspects of self-organization processes in composites. 2. The structure and interaction of inner boundaries. Nanomechanics Sci Technol **4**(2), 133–143 (2013)
18. Herega, A., Sukhanov, V., Vyrovoy, V.: The model of the long-range effect in solids: Evolution of structure clusters of interior boundaries, and their statistical descriptors. AIP Conf. Proc. **1909**, 020069 (2017)
19. Herega, A.N.: Physical aspects of self-organization processes in composites. 1. Simulation of percolation clusters of phases and of inner boundaries. Nanomechanics Sci Technol **4**(2), 119–132 (2013)
20. Herega, A., Ostapkevich, M.: Computer simulation mesostructure of cluster systems. AIP Conf. Proc. **1623**, 209–212 (2014)
21. Crownover, R.M.: Introduction to Fractals and Chaos. Jones and Barlett Publishers, London (1995)
22. Feder, J.: Fractals. Plenum Press, New York (1988)
23. Zel'dovich, Ya.B., Myshkis, A.D.: Elements of applied mathematics. Mir Publisher, Moscow (1976)
24. Zhukovskiy, N.Ye.: Teoreticheskaya mekhanika (Theoretical Mechanics). GITTL, Moscow (1952)
25. Plato: Republic. Hackett Publishing Company, Inc., Indianapolis (2004)
26. Mandelbrot, B.: The fractal geometry of nature. W.H. Freeman and Co., San Francisco (1982)
27. Bar'yakhtar, V.G., Gonchar, V.Y., Yanovskiy, V.V.: Priroda slozhnoy struktury pyatna zagryazneniy (The nature of the complex structure of the spots of contamination.). Ukraïns'kiy fizichniy zhurnal (Ukrainian Journal of Physics) **38**, 967–975 (1993)
28. Kuznetsov, V.R., Sabel'nikov, V.A.: Turbulence and combustion. Hemisphere Publishing Corporation, New York (1990)
29. Felk, E.V., Savin, A.V., Kuznetsov, A.P.: Effect of weak dissipation on the dynamics of multidimensional Hamiltonian systems. Nonlinear Phenom Complex Syst **18**(2), 259–265 (2015)

Chapter 6
Conclusion

One of the tendencies in the development of modern Natural Science and Mathematics is a spontaneous formation of specializations in sciences. The development of Materials Science is no exception.

In researches included in the monograph, the main attention is paid to the structure and interaction of the intermediate asymptotics components. Mesostructure of matter is a set of structural elements of different scale levels, for which spatial (statistical) self-similarity is characteristic—an example of multiscale structure. The experience of numerous experimental and theoretical researches on the study of mesostructure, the understanding of the physical body as a system, and structure as "a state arising as a result of the coherent behavior of a large number of particles" (G. Haken), allows us to assume that the most interesting results in the study of the condensed state of matter, and in particular, composite materials, will be obtained with consistent consideration of the interaction of structures of all scales, understanding of the aggregate of large-scale structures as a whole.

A. Herega, *The Selected Models of the Mesostructure of Composites*, SpringerBriefs in Physics, https://doi.org/10.1007/978-3-319-89704-2_6

Appendix A
Essay on Models

... They were thought out by a people.
Gennady Shpalikov, Russian poet

The classical definition of the model as an analogue of reality which is used to obtain and store knowledge about the original, its properties and structure, and also for the possibility of transforming and managing it [1, 2]—today requires detailed reflection and significant changes.

Models are among the main tools of cognition: the derivation of a theory, creating and studying of models are, in fact, the only ways to perform effective scientific research. The model is a means of the abstract description of some phenomena. What's more, if we assume that the model is a way to represent objects, events, and connections between them when they are absent, then we get a definition that essentially coincides with the description of the noetic consciousness [3] and emphasizes abstract, speculative, more precisely, intellectual nature of the model.[1]

Indeed, the concept of the model is an epistemological category. The possibility of modeling, i.e., the transfer of obtained results to the original, is based on the assumption that the model, in a sense, reflects and reproduces some features and characteristics of the object.

It is important to understand that the model should only describe cause-effect connections and properties of the object that are essential in a particular study. In describing the model, the researcher abstracts from a part of the object properties. He leaves only those that meet his interests, views, needs [4].

---

[1]Noesis (from the Greek "noetikos"—able to think) is an act of thought. In the philosophy of Plato is the mental discretion of the essence, based on the contemplative nature of the mind. According to E. Husserl is the semantic object, the objective meaning, which is the result of the active activity of consciousness.

A. Herega, *The Selected Models of the Mesostructure of Composites*, SpringerBriefs in Physics, https://doi.org/10.1007/978-3-319-89704-2

69

"Abstracting is, apparently, to get down to the point of matter, to get rid of the random features of the phenomenon and to focus on the most important properties" [5]. This makes it possible, in particular, to clarify the general structure of the investigated object, to reveal the essential cause–effect connections while separating the attributive features and characteristics from the secondary ones, which is quite difficult and is often the subject of a special study [6].

"However, the price which will have to pay for an unlimited abstraction is actually quite high. Unbounded abstraction is characterized by a tendency to divert attention from entire areas of application, and the very opening of which depends on those features that the abstract point of view excludes as random ones" [5]. Approximation of the model in this case can be either advantages or disadvantages. It depends on the level of education, experience, skill, and finally, success of the researcher, allowing him to pass between Scylla and Charybdis abstractions and details.

Models can also be understood as "a set of assumptions or postulates that describe some physical objects or phenomena...". Considering something as a model X means that there is some way of representing X, providing at least a certain approximation to the actual situation. Moreover, this means recognizing the possibility and necessity of alternative ways of representing X for other purposes [7].

Such a position echoes with the Niels Bohr complementarity principle [8], according to which for an adequate description of a physical object, it must be defined in mutually exclusive description systems [2] using additional classes of concepts [9]. The term "complementarity" emphasizes "the fact that... we are dealing with different but equally significant aspects of a single clearly defined set of information about objects" [8]. In particular, the principle focuses on the fact that clearly constructed logical systems act as metaphors. They define models that "behave in a sense the same way as the modeled system, and in some other way they do it differently" [9]. One logical construction is not enough to describe the complexity of the world picture [1, 2], and under these conditions the information deficit is compensated by its "stereoscopy": the opportunity to get a completely different product of the same reality, its translation into a completely different language [2].

Another aspect that characterizes the model and its cognitive function is related to Kurt Gödel's theorem on the incompleteness of deductive systems [10]. It follows from the theorem that the system is either not contradictory or incomplete, and that no fixed extension of this system axioms can make it complete: There will always be new truths not expressible by its means, and not derived from it.

As can be seen, the model by definition is attributively incomplete and, therefore, a priori contains contradictions. The creators of the model cannot claim its complete identity to the object. Rather, the model is a convergent set of polytypic analogies between the real phenomenon and the model ("Truth, as such, is in the process of cognition in its essence" [11–13]). And if we remember the simple restriction imposed on the model which can be considered either its definition (or part of it) or relevant convention: Nothing that is a model cannot be considered a model of itself, nor a model of something identical to itself [4], then from the question of the degree of model approximation to the original the features of surface perception are erased.

From a formal logical point of view, it is preferable to talk about models not as abstract representations of objects or phenomena, but as embodiments of a formal calculus when there are homomorphism relations between the structure of the formal system and the structure of its interpretation [4].

Model relations between some entities are triadic: "any entity can be considered as a model of any other if and only if we can isolate for them the general relevant properties—those by which one entity is like the other one" [4]. In accordance with this, a triad of the model relation can be formulated

$$M(S, x, y) \, \& \, R(x) < R(y), \tag{A.1}$$

i.e., subject S considers x as a model of y and where $R(x)$ and $R(y)$—areas of relevant properties [4]. As can be seen from the relation, the asymmetry of model relations arises because researchers do not consider models as rich in meaningful properties as the modeled object. Thus, believing that these objects have not yet known properties, and therefore "the specific feature of fruitful scientific models is the possibility of opening with their help those properties of simulated objects which we do not yet know" [4].

The material analogy, as is known, is not critical for the functional validity of the model. This analogy is "akin to the obvious validity: it guarantees nothing" [14]. "Any phenomenon is not 'just a thing', but a consequence of causal relationships between levels of factors in the real world" [14]. Therefore, any potentially useful model should use primary and formal analogies, known and explicated links. What's more, it should include a large number of causally related factors that may have obvious similarities or differ significantly and also establish the correspondence between them in various fields [14].

Essential for modeling is the approach to a set of accepted assumptions: for models they are considered as "useful" or "approximate" and do not express an explicit cognitive orientation toward truth [4]. Nevertheless, the "existential claims" [15] of models can vary significantly—from "ad hoc" to a situation where there is no "any plausible refutation of their adequacy" [15].

The attribute of models should be practicality and operability: paraphrasing the statement of Academician L. D. Landau, we can say that the true, and therefore, cumbersome, multi-formal or verbose description of the process, object, phenomenon, "can only be contemplated" [16]. The variety of tasks, which are solved by researchers, initiates the creation of models that differ in type, degree of fundamentalism, level of detail, measure of heuristics and universality, rigor of formalism, efficiency of algorithms, and other properties.

One of the groups of models is mathematical. The meaning of mathematical modeling is largely determined by the translation of the physical world into the language of mathematics, "which allows us to obtain a more accurate idea of their most essential properties" [17]. Such models reduce real circumstances by "sensibly simplifying the insignificant features of the situation, reducing the complex initial problem to an idealized one that can be mathematically analyzed" [17].

"Mathematical language is surprisingly well suited for the formulation of physical laws. This is a wonderful gift that we do not understand and do not deserve" [18]. "The incomprehensible effectiveness of mathematics in the natural sciences" [18] leads to the fact that mathematical equations often turn out to be effective models that enable by their kind and obtained solutions (analytical and numerical) to understand the behavior of the system [14]. These include, for example, stability studies, including the construction of characteristics stability regions in the parameter space, the identification of typical modes of functioning and evaluation of their attainability, analysis of asymptotic behavior, controllability and observability of systems [19].

The same system can be described by a variety of mathematical models that are eligible for the simultaneous existence, give results that are close to the experimental ones, and only "for reasons of a very general nature we have to choose the one which is of greater heuristic power" [9]. The "saving of thinking and experimentation" achieved at their use "is a consequence of the ease with which qualitative, seemingly reasonable working hypotheses can be excluded with the help of mathematical arguments" [20]. And "if before the description of real phenomena in mathematical language was perceived as an unconditional law of nature" [9], now it became possible to perceive them precisely as models.

The productive mathematical model also makes it possible to produce the necessary deduction and helps the researcher to obtain unexpected results that are impossible only on the basis of "common sense" [21]. It is interesting that sometimes "models based on clearly incorrect assumptions (e.g., the study of matter based on the idea of continuity of media) are productive in describing many phenomena" [6].

Several aspects make qualitative (non-calculated) models also irreplaceable. Firstly, a successful approach to the problem study, a correct allocation of the system, a formulation of basic provisions, etc., is an essential, sometimes even defining part of the study. Secondly, the qualitative theory has an ideological aspect which is not only interesting, but also useful. Thirdly, if the theory does not allow us to calculate the effect, then, it is possible, it will allow us to predict it, to specify conditions under which it can be observed, as well as accompanying phenomena, necessary and sufficient conditions, etc. Moreover, in relation to complex phenomena, even an "explanation in principle" with the help of qualitative models is naturally preferable to its absence [21].

Another way to study complex systems is simulation modeling.

The idea of simulation modeling is intuitively attractive due to its simplicity. They are convenient to use when you can not complete mathematical formulation of the problem or analytical methods of solution have not been developed or they are complex and time consuming. It can be the only one solution in a situation when it is impossible to set up a real experiment or to carry out field observations. In addition, simulation modeling makes it possible to fully control the time during which the process is carried out, which is both important and convenient [22].

Simulation models are unable to form solutions in such a form as analytical ones: "runs" are used to obtain necessary information with its help [22]. On the recommendation of Ernst Mach, having drawn up "a definite conclusion on the basis of one particular case, it is necessary gradually and as widely as possible to modify circumstances accompanying it, trying, as far as possible, to remain at the initial conclusion. There is no other way which, with great reliability and less mental effort, would lead to the simplest explanation of all the phenomena of nature" [23]. But before checking the correctness of the model, even before comparing with experimental data, one can be convinced of the unacceptability of the model on the analysis of simple limiting cases [24]. Such verification is necessary but not sufficient condition for the suitability of the model.

In the simulation experiment process, the model is both a means and an object of research replacing the original [1]. This suggests that the simulation modeling is not so much a theory of the phenomenon or process as problem-solving methodology for its study. And the simulation model itself is a logical–mathematical representation of the system [25]. Such modeling allows us to establish trends in the development of phenomena, to indicate ways to find the correct solutions already out of the model continuation.

When creating a simulation model, it is important to determine the level of accuracy of the real process description, i.e., choose a level of detailing that depends on the goal set by the researcher [26] and can lead to the hierarchy construction of object models [27], what's more, to determine the degree of reliability of information received.

The question of adequacy is important for any type of model, but in the case of imitation it is especially interesting. According to [25], the answer may well be: An adequate model with a certain degree of approximation mathematically and logically reflects the modeled system, process, principle or phenomenon; there is a correspondence between the logical elements of the model and the real system; probabilistic characteristics of the model reflect the nature of real random events; the outputs of the model is informative. Moreover, it is necessary to evaluate the adequacy of both the conceptual model and its implementation.

The simulation model does not reflect the real state of things. Thus, if we do not take this into account, the imitation of phenomena can lead to a wrong solution. The simulation model is in principle inaccurate. The degree of its accuracy cannot be measured, but we can only talk about the degree of its sensitivity to changing parameters.

On what, in this case, the sense of success in moving toward the goal is based.

Mental efforts have a heuristic effect: They seek to assimilate and use any suitable object in order to achieve the result [28]. When efforts are subordinated to the goal achievement, even the usual path of trial and error can lead to the necessary correction of further steps sometimes only partially realized.

Mastery of the cognition does not give a comprehensive description; a sense of success in moving toward the goal, if it accompanies our work, is explained not only by the stock of information, qualifications, and skill: Such confidence is the result of the personal knowledge, which, usually, manifests itself in the logical non-compliance of the action [28]. Indescribable in the form of rational formulations and textual constructions, the personal knowledge expresses the essence of the researcher's work. Furthermore, it is present in science as a style of a scientist, and most clearly manifested in the acts of discovery and their assessments by the scientific community.

It is difficult to determine the cause of success as a certain thing. This explains the wide dissemination of arts, skills, crafts that emerged as practical skills or occasional luck, and still constitute an important part of technology that is rarely fully specified [28].

References

1. Filosofskiy entsiklopedicheskiy slovar'. (Philosophical Encyclopedic Dictionary). Sovetskaya entsiklopediya, Moscow (1989)
2. Levita, S.Ya. (ed.): Entsiklopediya «Kul'turologiya. XX vek». (Encyclopedia "Culturology. XX Century"), vol. 2. Universitetskaya kniga, Sankt-Peterburg (1998)
3. Solso, R.L.: Cognitive Psychology. Harcourt-Brace, New York (1979)
4. Wartofsky, M.W.: Models. Representation and the Scientific Understanding. D. Reidel Publishing Company, London (1979)
5. Kac, M., Ulam, S.: Mathematics and Logic: Retrospect and Prospects. F.A. Prager Inc., New York (1968)
6. Sedov, L.I.: Similarity and Dimensional Methods in Mechanics. CRC Press, New York (1993)
7. Rapoport, A.: Mathematical aspects of general systems analysis. Gen. Syst. XI, 3–25 (1966)
8. Bohr, N.: Can quantum-mechanical description of physical reality be considered complete? Phys. Rev. 48(8), 697–702 (1935)
9. Nalimov, V.V.: Veroyatnostnaya model' yazyka. (The Probabilistic Model of Language). Nauka, Moscow (1979)
10. Uspenskiy, V.A.: Teorema Godelya o nepolnote. (Gödel's Incompleteness Theorem). Nauka, Moscow (1982)
11. Hegel, G.: The Science of Logic. Cambridge University Press, New York (2010)
12. Russell, B.: Human Knowledge. Its Scope and Limits. George Allen & amp; Unwin Ltd., London (1948)
13. Solov'yov, V.S.: Sochineniya, vol. 2 (Collected works). Mysl', Moscow (1988)
14. Corsini, R., Auerbach, A.: Concise Encyclopedia of Psychology. Wiley, New York (1996)
15. Achinstein, P.: Models. Philos. Sci. 31, 328–350 (1964)
16. Migdal, A.B.: Kachestvennyye metody v kvantovoy teorii. (Qualitative Methods in Quantum Theory). Nauka, Moscow (1975)
17. Andrews, J.G., McLone, R.R. (eds.): Mathematical Modeling. Butterworths, London-Boston (1976)
18. Wigner, E.P.: Symmetries and Reflections. Indiana University Press, Bloomington-London (1970)
19. Matematicheskiy entsiklopedicheskiy slovar'. (Mathematical Encyclopedic Dictionary). Sovetskaya entsiklopediya, Moscow (1988)
20. Feller, W.: An Introduction to Probability Theory and Its Applications, vols. 1, 2. Wiley, New York (1970)
21. von Bertalanffy, L.: General System Theory. A Critical Review. General Systems, vol. VII (1962)

22. Shannon, R.E.: Systems Simulation. The Art and Science. Alabama Prentice-Hall Inc., New Jersey (1975)
23. Mach, E.: The Science of Mechanics: A critical and Historical Account of Its Development. Open Court Press, Chicago-La Salle (1960)
24. Weyl, H.: Philosophy of Mathematics and Natural Science. Princeton University Press, Princeton (1950)
25. Martin, F.: Computer Modeling and Simulation. Wiley, New York (1968)
26. Tomashevs'kiy, V.M.: Modelyuvannya sistem. (Simulation of Systems). BHV, Kiev (2005)
27. Zarubin, V.S.: Matematicheskoye modelirovaniye v tekhnike. (Mathematical Modeling in Engineering). Izdatel'stvo MGTU im. N. E. Baumana, Moscow (2003)
28. Polanyi, M.: Personal Knowledge: Towards a Post-critical Philosophy. University of Chicago Press, Chicago (1958)

Appendix B
Some Parameters of Percolation Systems. Calculating Formulas

Notation:

N—the number of elements of the percolation field, or the number of elements in the field with an edge L;

s—the number of elements in the final cluster (the mass of the final cluster);

N_S—the number of clusters containing s elements;

D—fractal dimension;

d—dimensionality of space.

Calculating formulas:

$$p = \frac{N_{cl.}}{N} \tag{B.1}$$

– fraction of cluster system elements;

$$n_s(p) = \frac{N_S}{N} \tag{B.2}$$

– average (for one point) number of finite clusters containing s elements;

$$\sum_s s \cdot n_s(p) \tag{B.3}$$

– fraction of elements occupied by all finite clusters;

$$\omega_s = \frac{s \cdot n_s}{\sum_s s \cdot n_s} \tag{B.4}$$

A. Herega, *The Selected Models of the Mesostructure of Composites*, SpringerBriefs in Physics, https://doi.org/10.1007/978-3-319-89704-2

– probability that an element belongs to a cluster containing s elements;

$$r_0 = \frac{1}{s}\sum_s r_i \tag{B.5}$$

– radius-vector of the cluster's center of mass;

$$\langle s \rangle = \sum_s \omega_s \cdot s = \frac{\sum_s s^2 \cdot n_s}{\sum_s s \cdot n_s} \tag{B.6}$$

– average mass of the final cluster;

$$R^2(s) = \frac{1}{2s^2}\sum_{i,j=1}^{s}[(r_i - r_j)^2] \tag{B.7}$$

– square of the radius of isotropic clusters gyration;

$$R^2(s) = \frac{1}{s}\sum_{i,j=1}^{s}[(x_{m,i} - x_{m,j})(x_{n,i} - x_{n,j})], \tag{B.8}$$

where $m, n = 1, 2$

– square of the radius of anisotropic clusters gyration;

$$A = \frac{R_{11}}{R_{22}} \tag{B.9}$$

– anisotropy degree;

$$s \sim R^D \tag{B.10}$$

– relation between the number of elements in the cluster and the radius of gyration;

$$s \sim L_s^D \tag{B.11}$$

- relation between the number of elements in the cluster and the edge of the minimal cell into which the cluster is inscribed;

$$n_s \sim s^{-\tau} \tag{B.12}$$

where $\tau = \dfrac{d+D}{D} = 1 + \dfrac{d}{D}$

- power growth index;

$$\langle s \rangle^2 - \langle s \rangle^2 \sim L_s^{2D} \tag{B.13}$$

- lacunarity: fluctuations in the value of s at a given value L_s;

$$\langle s \rangle \sim L^{2D-d} \tag{B.14}$$

- average cluster mass;

$$P_\infty(p) = p - \sum_s s \cdot n_s(p) = \frac{N_{\text{inf.}}}{N} \tag{B.15}$$

- power of an infinite cluster;

$$\langle R^2(s) \rangle_L = \frac{\sum_s \langle R^2(s) \rangle \cdot s \cdot n_s}{\sum_s s \cdot n_s} \sim L^{2-(d-D)} \tag{B.16}$$

- average value of the radius of gyration;

$$\xi^2 = \frac{2 \sum_s R^2(s) s^2 n_s}{\sum_s s^2 n_s} \tag{B.17}$$

- correlation length.

Appendix C
Dimensions: Genesis of Representations and Physical Applications

Introduction. The dimension theory, the basis of which are known to have been laid down in researches of Poincaré, Lebesgue, and Brouwer [1–3] published during the period from 1911 to 1913, has opened "access to the study of a number of interesting properties of point sets and to the formulation of the extensive theory ..." [4, 5]. The notion of the dimension proved to be useful and convenient concept in Theoretical Physics as well.

Two basic groups of problems, in which the dimension has become a real tool for obtaining and analyzing the solutions, have formed in Physics. The first group contains problems of stochastic dynamics, in particular, problems of turbulence. For these problems, dimensions are available for measuring and they are structurally stable characteristics of the system, associated in particular with Lyapunov indicators; allow classification of strange attractors and the associated with them chaotic behavior [6, 7].

The second group of problems contains problems of percolation theory: the section of Statistical Physics, which for half a century is studying the critical phenomena [6, 7]. The percolation theory adequately describes the features of the origin and the evolution, as well as properties of the connected areas in systems with the geometrical phase transition. It has been applied in a wide range of scientific and technical problems: the research of protein structures, porous bodies, the creation of filters, the research of doped semiconductors, at the struggle with epidemics, the research of polymerization processes, the development of composite materials, the study of philosophical questions, and many others. At the same time for critical indicators of physical quantities describing the processes and phenomena, as a rule, a set with the dimension to which this indicator is linked can be specified. In its turn, the structure research of these sets gives a lot to the understanding of the system critical behavior and of relations between indicators, allows tracking the connection between the behavior of the system in the intermediate asymptotic and its geometry [6, 7].

© The Author(s), under exclusive licence to Springer International Publishing, part of Springer Nature 2018
A. Herega, *The Selected Models of the Mesostructure of Composites*, SpringerBriefs in Physics, https://doi.org/10.1007/978-3-319-89704-2

81

The percolation theory is also known as the section of the probability theory, which has its own application in natural sciences and engineering [8–11].

Mathematical and physical researches use a large number of dimensions. Each particular research always raises the issue of choice which is caused by the content of the problem. Dimensions shown in the survey relate to quite different concepts, in particular, they relate to described group of tasks (primarily to the percolation theory), and to topological studies as well.

Further in the text, as in all physical applications, dimensions are defined as the exponent in expressions such as $a \sim b^c$.

Geometrical and physical dimensions. The dimension of the physical quantity determines its relation with quantities underlying the system of measurement units, i.e., sets the ratio of scales of given and basic units of measurement [12]. This definition implies that physical quantities, in contrast to geometrical, vary not only with coordinate transformations, but with the modification of the system of physical quantities. It means that at a certain choice of basic units one and only one geometric object corresponds to each physical object—a image of the physical object, which acquires a factor $a^{\alpha}\ \beta^{b}\ c^{\gamma}\ ...\ l^{\lambda}\ m^{\mu}\ n^{\nu}$, when applying a different set of admissible basic units. This factor is called the dimension of the geometric image or the absolute dimension of the physical object. The absolute dimension is defined relatively to the affine group, if we are working with rectilinear coordinates of the n-dimensional Euclidean space. This dimension is not the one that works with Physics [13].

If components of the physical object relatively to the local Cartesian coordinate system based on the unit length, when changing the basic units gain the factor $a^{\alpha 1}\ \beta^{b1}\ c^{\gamma 1}\ ...\ l^{\lambda 1}\ m^{\mu 1}\ n^{\nu 1}$, then it is called the relative dimension or simply the dimension of the physical object. This is the one that is used in Physics.

Thus, the difference between the absolute (geometric) and the relative (physical) dimension of the physical object caused by the fact that the common coordinate system is linked to unit length and the local Cartesian coordinate system changes when applying the another unit. Hence, the ratio between them depends only on the law of transformation of the object during the coordinate transformation [13].

Dimension of the space. Topological dimension. If one perceives geometric objects as a set of points in Euclidean space R^{E}, then the notion of the topological dimension can be presented in the recurrence scheme proposed by Poincaré [1].

Suppose that the dimension of any finite or countable set of points is equal to zero; the dimension of connected set will be considered equal to $d + 1$ if it can be cut into two unconnected parts by excluding from it at least d-dimensional set of points, i.e., performing a d-dimensional incision. Under this definition, if we put that the dimension of a point is equal to zero, then the topological dimension of a line will be equal to one, the topological dimension of a plane and a sphere will be equal to two, ball—three and so on.

From the definition, it is clear that the topological dimension can only be an integer number and coincides with the intuitive idea of a minimum number of variables that must be set to determine the position of the point at the object [14]. In 1902, in his book "Science and Hypothesis" [15], Poincaré wrote that "the dimension of the space is the minimum number of parameters that is required in order to differentiate points of the space from each other"; a complete visual space "has three dimensions; i.e. elements of our visual sensations ... will be completely determined when we know three of them". What's more, if in the space this number equals to three, on the plane only need two coordinates, on the line—one; in this sense, the space is three-dimensional, the plane is two-dimensional, the line is one-dimensional.

In the article "Why does the space have three dimensions" [1] H. Poincaré defined the dimension, which could be called anthropic. He writes about the existence of "experimental facts that make us attribute three dimensions to the space". Given this data, it was convenient to attribute three dimensions to it, instead of four or two. But the word "convenient", perhaps, in this case is not strong enough: "a creature that would be attributed two or four dimensions to the space, would have been less adapted to the struggle for existence in a world like ours". In case of two dimensions, it would presuppose the existence of such ratio that we, human beings, do not allow; and in case of four—would discard ratio that we use [1].

Physical justification for the three-dimensional space is given in study of Ehrenfest [16], where, in particular, the analogue of Newton's law of gravity for spaces with different number of dimensions is investigated. In this case, dependence of gravitational forces on the distance determined by the expression $F = G_i M m / R^{i-1}$, where M, m, R—masses and the distance in "classical" sense, i—the number of spatial coordinates, G_i—coefficients, particularly, G_3—Newtonian gravitational constant. As some grounds for this assumption are the results of the analysis of motion laws which Ehrenfest received for spaces with the number of dimensions different from three [16, 17].

In the gravitation law proposed by Newton, force is proportional to R^{-2}. Moreover, this dependence was repeatedly verified experimentally, and the value of the exponent was established up to $2 \pm 3 \cdot 10^{-11}$; hence, the same accuracy has the dimension of our space $i = 3$. This data was derived from precise measurements of the orbit of the Moon moving around the Earth. And, although, the average radius of the lunar orbit is 384,000 km, the model data is different from the measured ones on 4 mm [18].

Adequacy of measure. Hausdorff-Besicovitch dimension. The notion of the topological dimension decides the question of an adequate measure of classical geometrical objects. A simple way to measure the length of curves, area of surfaces and volumes of bodies consists in dividing the space into small cubes or spheres with a characteristic linear dimension δ. Counting the number of line segments, squares, or cubes required to cover a given set of points, it is possible to get the measure of this set [19].

For a typical curve, the length L can be determined by a limit passage

$$L = N(\delta)\delta \xrightarrow[\delta \to 0]{} L_0 \delta^0, \tag{C.1}$$

where δ—the length of straight line segments, $N(\delta)$—their number. As seen, in the limit as $\delta \to 0$, the measure L becomes asymptotically equal to the length of the curve and it does not depend on δ.

The length is an adequate measure of the ordinary (non-fractal) curve. If put in correspondence with line not the length but the surface area or the volume, then the same procedure will reveal that such measures become zero. Indeed, let $N(\delta)$—the amount of squares required to cover the curve, δ^2—the area of one square, then

$$S = N(\delta)\delta^2 \xrightarrow[\delta \to 0]{} L_0 \delta^1 = 0; \tag{C.2}$$

similarly,

$$V = N(\delta)\delta^3 \xrightarrow[\delta \to 0]{} L_0 \delta^2 = 0. \tag{C.3}$$

The centerpiece of the definition of the Hausdorff–Besicovitch dimension takes the concept of the measure adequacy.

Consider the set of points forming a surface in three-dimensional space. An adequate measure of such a set is the area. Indeed,

$$S = N(\delta)\delta^2 \xrightarrow[\delta \to 0]{} S_0 \delta^0, \tag{C.4}$$

where S_0—surface area. Thus, the amount of squares required to cover the surface is determined in the limit when $\delta \to 0$ as $N(\delta) = S_0/\delta^2$.

If put in correspondence with the surface the length, then

$$L = N(\delta)\delta \xrightarrow[\delta \to 0]{} S_0 \delta^{-1} = \infty, \tag{C.5}$$

that is, indicating that it is impossible to cover the surface with a finite number of straight line segments. If make an attempt to establish a correspondence between the surface and volume, then it becomes zero.

$$V = N(\delta)\delta^3 \xrightarrow[\delta \to 0]{} S_0 \delta^1 = 0. \tag{C.6}$$

Define the measure of a fractal set using the trial function M_d which depending on the choice of its dimension d would become equal to zero or infinity when $\delta \to$

0. Let us introduce the Hausdorff–Besicovitch dimension D_H, where the measure M_d changes the value from zero to infinity

$$M_d = \sum \gamma(d)\delta^d = \gamma(d)N(\delta)\delta^d \underset{\delta \to 0}{\to} \begin{cases} 0, d > D \\ \infty, d < D \end{cases} \tag{C.7}$$

where $\gamma(d)$—a geometric coefficient that depends on the shape of the elements, covering the set; for squares and cubes $\gamma(d)$ is equal to one, for circles—$\pi/4$, for spheres—$\pi/6$ [15]. For a physicist such behavior of the measure means that D represents a critical dimension [20].

Essential that in the determination of the Hausdorff–Besicovitch dimension, it is necessary to cover a lot of elements of various sizes that do not exceed a certain small value, and to determine the infimum expression $\gamma(d)\sum \delta^d$. Obviously, the process of minimizing of this amount over all possible partitions is extremely laborious. And it is usually produced an assessment of the Hausdorff–Besicovitch dimension by value of capacity of the set D_c. This is a typical situation in the applied dimension theory: Among the same type of dimensions, it is sometimes possible to find ones that are useful for estimation the values of others, the calculation of which is laborious or unrealizable.

In order to determine D_c consider the case of coating the set of points in d-dimensional Euclidean space by a minimum amount of d-dimensional cubes (or spheres) of the same size. (Coverage by spheres is used in order not to speak of orientation). That is, if $N(\delta) \sim \delta^{-D_c}$, then D_c—Kolmogorov sets capacity [6, 21]; index c—an abbreviation of 'capacity'.

Suppose that A—the value characterizing the coverage, and $N(\delta) \approx A \cdot \delta^{-d}$—the minimum number of d-dimensional cubes. Taking the logarithm of this expression, we obtain

$$\ln N(\delta) \approx \ln A - d \cdot \ln \delta, \tag{C.8}$$

from which, approximately,

$$d = -\frac{\ln N(\delta)}{\ln \delta} + \frac{\ln A}{\ln \delta}. \tag{C.9}$$

In as much as $\ln \delta \to -\infty$ when $\delta \to +0$, then the sets capacity is the limit

$$D_c = -\lim_{\delta \to 0} \frac{\ln N(\delta)}{\ln \delta}, \tag{C.10}$$

which normally exists.

Since when determining the Hausdorff dimension, all kinds of coverages of the set must be used and when calculating the capacity—elements of the same size, then

$$D_H < D_c. \tag{C.11}$$

The Hausdorff dimension and the Kolmogorov capacity may vary for even very simple sets [6]. For an instance, for a set of points of the straight line with coordinates $x_n = 1/n$, the former is equal to 0, the latter—1/2. Capacities unlike dimensions, in particular the Hausdorff one, do not remain invariant under the piecewise smooth, possibly with features, transformation of coordinates, and for the value determined as the dimension such invariance is needed.

As mentioned before, in all physical applications the dimension is defined as the coefficient $M \sim l^D$, where M—some property, l—a characteristic size, and it is not possible to determine whether it is the Hausdorff dimension or capacity. This is due to the fact that the dimension describes properties of the intermediate asymptotic behavior and the transition to the limit, required by the formal definition, is impossible. In addition, at small scales the system is not fractal, and its behavior is described by some minimum scale.

So, the Hausdorff–Besicovitch dimension and the capacity, that were defined, are local properties in the sense that they characterize the set of points with a vanishingly small characteristic dimension of the trial function ($\delta \to 0$), used to cover it.

It is important that for simple geometric objects, the Hausdorff dimension coincides with the topological one. Indeed, suppose that there is a square with the side a; let us cover it with small squares of the area of δ^2. Then, for the measure $A_d(\delta)$ we obtain

$$A_d(\delta) = \sum \delta_i^d = N(\delta) \cdot \delta^d \approx a^2 \delta^{d-2}. \tag{C.12}$$

When $d < 2$, the measure $A_d(\delta)$ increases indefinitely with the decreasing of δ. If $d > 2$—it tends to zero. Therefore, by definition of the Hausdorff–Besicovitch dimension $D_H = 2$.

Hausdorff–Besicovitch dimension as a fractal dimension. In Euclidean space, R^E the value of the topological dimension D_T and the Hausdorff-Besicovitch dimension D_H are between zero and E. Herewith, the topological dimension is always integer, whereas, for the Hausdorff–Besicovitch dimension it is not necessarily so. For the Euclidean sets $D_H = D_T$, in general, these two dimensions must satisfy the inequality of Edward Szpilrajn $D_H \leq D_T$ [5, 20].

However, there are sets for which $D_H > D_T$. In [20], B. Mandelbrot writes: "These sets had to be called somehow, so I coined the term "fractal", defining it as follows: The fractal is a set whose Hausdorff–Besicovitch dimension is strictly greater than its topological dimension".

Any set with non-integer value of D_H is a fractal; the fractal may have the integer dimension. If understand the term "fraction" as a synonym for the expression "non-integer real number", then often values of the dimension D_H are fractional. Given that D_H can also take integer values $D_T < D_H \leq E$, Benoit Mandelbrot preferred to call it the fractal dimension [20] and denoted it as D.

Variety of coverages. Minkowski–Bouligand dimension and Pontryagin–Schnirelman dimension. According to Mandelbrot, the fractal dimension and all its possible options are not topological but metrical notions: Each one includes a metric space, where distances between any two points are defined [20]. Herewith, dimensions themselves are determined by algorithms of coverage of the set by d-dimensional balls and, in fact, there are functions of a covering method.

In the covering method of a limited set offered by G. Cantor, each point is regarded as a center of a ball [20]. This approach is connected to the obvious discomfort. Firstly, for sets containing non-finite number of points, such an algorithm is inoperable. The situation, however, is being solved because, it appears, that it is enough to construct a finite number of balls $N(\rho) \sim 1/\rho$.

Secondly, a sum of overlapping volumes of balls when $\rho \to 0$ does not necessarily converge to the extent (in the sense of Minkowski) of the set, i.e., to the d-dimensional volume: length, area, volume, etc. In the example of Schwarz [22], it is shown that with an increase of the accuracy of triangulation of the lateral surface of a right circular cylinder, a sum of triangles areas does not necessarily converge to its area and can be equal to an arbitrarily large finite or infinite value.

When considering this paradox, H. Minkowski demonstrated that if define the length as

$$V\{d - \text{dimensional ball of radius } \rho\} = \gamma(d)\,\rho^D, \qquad (C.13)$$

where d—the standard topological dimension of the considered set, the factor $\gamma(d) = [\Gamma(\frac{1}{2})]^d/\Gamma(1 + d/2)$, ρ—the radius of covering balls, D—the dimension of the fractal set (if the set is non-fractal, $D = d$), then when $\rho \to 0$ it cannot have a limit [20]. In this case, the expression $\lim\limits_{\rho \to 0} V$ is replaced by $\lim\limits_{\rho \to 0} \sup V$ and $\lim\limits_{\rho \to 0} \inf V$—upper and bottom extents of the set. At the same time, for every real number from the interval $]\lim \inf, \lim \sup[$ there is at least one sequence of values $\rho_m \to 0$ such that when $m \to \infty$ the sum of triangles areas in the example of H. A. Schwarz converges to the surface area [20]. H. Minkowski also showed that in case of the standard Euclidean structures, there is a value D_M—the Minkowski dimension—such that if $d > D_M$, then the upper extent of the set becomes zero, and when $d < D_M$ the bottom one is infinite [20].

In 1928, J. Bouligand generalized the Minkowski dimension for the case of fractional d and showed that it is defined by the expression $\lim\limits_{\rho \to 0} \inf V$. In some cases, values of the Minkowski–Bouligand dimension D_{MB} and the Hausdorff-Besicovitch dimension D_H resemble closely, for example, for smooth curves and surfaces. Given also that D_{MB} is easier to estimate and, by analogy with the situation with the Hausdorff–Besicovitch dimension and capacity, it can be used for definition of value D_{H} [20].

It is not difficult to give an opposite example, when such an estimation is not possible: For the compact set $\{0, 1, 1/2, 1/3, 1/4, \ldots\}$ as for any countable, $D_{\mathrm{H}} = 0$ and $D_{\mathrm{MB}} = \frac{1}{2}$. In general, as shown in [23] $D_{\mathrm{MB}} \geq D_{\mathrm{H}}$.

Among all the possible sets of covering balls, the most economical is a package containing at least $N(\rho)$ balls, which is used to determine the Pontryagin–Schnirelman dimension [20, 24]

$$D_{\mathrm{PSch}} = \lim_{\rho \to 0} \inf \ln N(\rho)/\ln(1/\rho). \qquad (C.14)$$

Self-similarity dimension and cellular dimension. As it is known, the objects that are zooming-invariant and parallel transfer-invariant are called self-similar. If zooming in n times ($n < 1$) could let cover the initial object by reduced copies once, then this object is self-similar with a similarity factor $n\,(N) = 1/N^{1/d}$, where N—the number of identical parts with n times smaller linear size, d—the similarity dimension (self-similarity) which is equal to the topological dimension of the object. In case of geometric self-similar (regular) fractals,

$$n(N) = 1/N^{1/D_{\mathrm{S}}} \qquad (C.15)$$

where D_{S} coincides with the Hausdorff–Besicovitch dimension [19] and is determined by the formula

$$D_{\mathrm{S}} = \ln N / \ln n. \qquad (C.16)$$

As an illustration, consider a square divided into $N = 4$ equal squares with sides of $n = 2$ times smaller than the original one. Then, the self-similarity dimension, which is equal to the topological one, for the square possesses value of 2; for the cube divided into $N = 8$ equal pieces $D_{\mathrm{S}} = \ln 8/\ln 2 = 3$. For the Koch curve [19]— the regular fractal, in the construction of which the scale of covering segments is reduced to $n = 3$ times at each iteration, and their number becomes equal to $N = 4$, the dimension $D_{\mathrm{S}} = \ln 4/\ln 3 = 1.2618\ldots = D_{\mathrm{H}}$.

To determine the dimension of irregular objects of the fractal type, for example, images of a state border or a coastline, described algorithm, naturally, is not suitable. Therefore, another definition of the dimension connected with a different algorithm is used.

The space which contains the object of interest is divided into cells with the size δ^2. For an instance, to the object image a square grid with the side δ is applied, and then the number of cells containing the object points is counted. Fragmentation is repeated multiple times using smaller and smaller scale. Dependence of the number of cells which have object points from δ is described as $N(\delta) = A\delta^{D_{\mathrm{S}}}$, where D_{S}—

the desired fractal dimension of self-similarity. To calculate its value a dependency graph of $N(\delta)$ in the double logarithmic scale is built, where the value of D_S is defined by the slope ratio.

For the dimension defined by such algorithm, a title of the self-similarity dimension could be preserved: Indeed, if apply the described algorithm to regular objects, values of dimensions coincide with those calculated according to the formula for the D_S. However, as the dimension defined by counting the number of cells it is called the cellular dimension [6, 19].

Dimension of entropy of measure. The described algorithm for calculating the dimension of the irregular fractal set has a natural limitation: If, for example, a coastline is heavily indented and it repeatedly crosses some cell, then to the number of cells covering the set of points it still provides a unit contribution that is "not completely fair" [19].

Let us consider the distribution of set points among the cells, which reflects the distribution of measure. Let the set of N points contains in the ith cell N_i points, $\mu = N_i/N$—the probability of the cell filling. It is possible to construct the measure

$$M_d(q,\delta) = \sum_{i=1}^{N} \mu_i^q \delta^d = N(q,\delta)\delta^d \xrightarrow[\delta \to 0]{} \begin{cases} 0, d > \tau(q) \\ \infty, d < \tau(q) \end{cases} \qquad (C.17)$$

with the exponent $d = \tau(q)$, at which it does not become zero or infinity when $\delta \to 0$ [19]. The measure is characterized by the whole sequence of exponents $\tau(q)$ that defines the degree law according to which probabilities $\{\mu\}$ are changing depending on δ. In this case, the weighted number of cells is equal to

$$N(q,d) = \sum_i \mu_i^q \sim \delta^{-\tau(q)}, \qquad (C.18)$$

where

$$\tau(q) = -\lim_{\delta \to 0} \frac{\ln N(q,\delta)}{\ln \delta}. \qquad (C.19)$$

From these relations, it is clear that when $q = 0$, we obtain

$$\mu_i^{q=0} = 1. \qquad (C.20)$$

Then, $N(q = 0, \delta)$—the number of points covering the set and $\tau(0) = D$—the fractal dimension of the set. Furthermore, considering, we obtain $\tau(1) = 0$.

Let us introduce the derivative

$$\frac{d\tau(q)}{dq} = -\lim_{\delta \to 0} \frac{\sum_i \mu_i^q \ln \mu_i}{\left(\sum_i \mu_i^q\right) \ln \delta}, \tag{C.21}$$

and consider

$$\frac{d\tau(q)}{dq}\bigg|_{q=1} = -\lim_{\delta \to 0} \frac{\sum_i \mu_i \ln \mu_i}{\ln \delta} = \lim_{\delta \to 0} \frac{S(\delta)}{\ln \delta}, \tag{C.22}$$

where $S(\delta)$—the information entropy of measure M partition among cells with the size δ, which can be expressed as

$$S(\delta) = -\sum_i \mu_i \ln \mu_i \sim -\alpha_1 \ln \delta. \tag{C.23}$$

The exponent $\alpha_1 = -\frac{d\tau(q)}{dq}\big|_{q=1}$ is the fractal dimension of the set, at which the measure is focused; it describes the scaling behavior of the entropy of measure partition when resizing cells δ [19]; up to the factor it is equal to the information dimension—the second one of the spectrum of generalized Rényi dimensions.

Generalized Rényi dimensions. For an adequate description of inhomogeneous fractal objects (multifractals), the spectrum of generalized A. Rényi fractal dimensions can be used [14, 25, 26].

Suppose there is a fractal object, limited by an arbitrary area with a size L in the Euclidean space with a dimension d. And let it be a set of $N \to \infty$ points randomly distributed in this area. Let us divide the whole area into rectangular cells with side $\delta \ll L$ and volume δ^d. Cells that contain at least one point are defined as occupied. Suppose $N(\delta)$—the total number of occupied cells, $n_i(\delta)$—the number of points in the ith cell; define the probability of location of an arbitrary point of the set in the I cell as

$$p_i(\delta) = \lim_{N \to \infty} [n_i(\delta)/N], \tag{C.24}$$

and introduce a generalized statistical sum

$$Z(q, \delta) = \sum_{i=1}^{N(\delta)} p_i^q(\delta). \tag{C.25}$$

Then, according to Renyi [25, 26], we can introduce the spectrum of generalized dimensions characterizing distribution of points in the arbitrary area and showing how it is heterogeneous. It is essential that these dimensions are related to exponents $\tau(q)$ described in the previous section as

$$D_q = \frac{\tau(q)}{q-1}, \tag{C.26}$$

where

$$\tau(q) = \lim_{\delta \to 0}(\ln Z(q, \delta)/\ln \delta), \tag{C.27}$$

and the factor is selected to ensure for sets of a constant density in d-dimensional space the equality $D_q = d$ [19].

Indeed, if for a uniformly distributed measure in d-dimensional space with a constant density of points, we divide the space into $N = \delta^{-d}$ cells with the volume δ^d, then $\mu_i = \delta^d$ and

$$\sum_{i=1}^{N} \mu_i^q = \sum_{i=1}^{N} \delta^{qd} = \delta^{(q-1)d}, \tag{C.28}$$

and, therefore,

$$D_q = \frac{1}{q-1} \lim_{\delta \to 0} \frac{\ln \delta^{(q-1)d}}{\ln \delta} = d. \tag{C.29}$$

Thus, the spectrum of fractal dimensions for the uniformly distributed measure is reduced to the dimension of space and does not depend on the order of the degree of the moment q [19].

Information dimension. Correlation dimension. Properties of function D_q.
Define the meaning of generalized Rényi dimensions D_q for $q = 1$ and $q = 2$ [14, 25, 26]. Generalized statistical sum by virtue of the condition of the probability normalization when $q = 1$ is equal to one which obviously leads to an uncertainty $0/0$ in the expression for D_1.

Clarify the uncertainty with the help of expression

$$Z(q, \delta) = \sum_{i=1}^{N(\delta)} p_i^q = \sum_{i=1}^{N(\delta)} p_i \exp[(q-1) \ln p_i]. \tag{C.30}$$

Letting $q \to \infty$ and decomposing an exponent into a series, taking into account the normalization condition, we obtain

$$Z(q \to 1, \delta) \approx \sum_{i=1}^{N(\delta)} [p_i + (q-1)p_i \ln p_i] = 1 + (q-1) \sum_{i=1}^{N(\delta)} p_i \ln p_i. \tag{C.31}$$

Then,

$$D_1 = \lim_{\delta \to 0} \sum_{i=1}^{N(\delta)} p_i \ln p_i / \ln \delta. \tag{C.32}$$

Up to a sign, the numerator in this formula is the entropy $S(\delta)$ of a fractal set

$$S(\delta) = -\sum_{i=1}^{N(\delta)} p_i \ln p_i, \tag{C.33}$$

and

$$D_1 = \lim_{\delta \to 0} \frac{S(\delta)}{\ln(\delta)}. \tag{C.34}$$

Thus, the value D_1 characterizes the information necessary to determine the position of a point in a certain cell and it shows how the amount of information increases when the cell size tends to zero $S(\delta) \approx \delta^{-D_1}$. It is called the information dimension [14, 25].

Definition of D_1 via entropy makes it possible to compare the values of fractal and information dimensions: The calculation of D_1 is meaningful only in the case of the inhomogeneity of the fractal set of points for which the entropy is less than for a homogeneous one, and, therefore, always $D_1 < D_0$ (equality is obviously possible only for the homogeneous case). In the dimension theory, this result is generalized for an arbitrary q, and it is shown that there is an inequality

$$D_0 \geq D_1 \geq D_2 \geq D_3 \ldots \tag{C.35}$$

To determine the physical meaning of the generalized dimension D_2 which is equal to

$$D_2 = \lim_{\delta \to 0} \left(\ln \sum_{i=1}^{N(\delta)} p_i^2 / \ln(\delta) \right), \tag{C.36}$$

introduce the paired correlation integral

$$I(\delta) = \lim_{N \to \infty} \frac{1}{N^2} \sum_{n,m} \theta(\delta - |r_n - r_m|). \tag{C.37}$$

Here the summation is carried out over all pairs of the fractal set's points with radius vectors r_n and r_m, and $\theta(x)$—Heaviside step function. This sum determines the number of pairs of points (n, m) located at a distance less than δ and, thus, being divided by N^2, it is equal to the probability that two arbitrary points have this property.

This probability could also be found otherwise [14, 25, 26]. By definition, the value p_i^2—probability of hitting of two points in the ith cell with a size of δ. Summing p_i^2 over all occupied cells, we obtain the probability that two randomly selected points lie inside of the plurality of single cells with the size of δ. Consequently, the distance between them will be approximately δ or less. Then, up to numerical coefficients

$$I(\delta) \approx \sum_{i=1}^{N(\delta)} p_i^2 \approx \delta^{D_2}. \tag{C.38}$$

In this way, the generalized dimension D_2 determines a dependence of the correlation integral $I(\delta)$ in a limit $\delta \to 0$ and it is called correlation [25].

Using the probabilistic interpretation, it is convenient to clarify the meaning of the boundary dimensions of the Renyi spectrum: minimum

$$D_\infty = \lim_{q \to \infty} D_q \tag{C.39}$$

and maximum

$$D_{-\infty} = \lim_{q \to -\infty} D_q. \tag{C.40}$$

The value $D_{-\infty}$ defines the formal upper boundary of the interval of changes D_q (maximum value is of the dimension D_0, since dimensions in the spectrum are defined for nonnegative q), the dimension D_∞—minimal among dimensions. When $q \to \infty$, the main contribution to the generalized partition function is obviously made by cells with the highest probability of filling, i.e., containing the maximum number of particles; when $q \to -\infty$, respectively—the most sparse.

Then, in view of the above,

$$D_{-\infty} \geq D_0 \geq D_1 \geq D_2 \geq D_3 \ldots \geq D_\infty. \tag{C.41}$$

As an illustration, let us consider the calculation of the spectrum of Renyi dimensions for the inhomogeneous Cantor set [27].

Suppose that two intervals, remaining from the segment formed in the previous step of the iterative procedure for constructing the set, have lengths that are related to as a/b, and $a + b = 1$. Then, the probability appearing in the expression for the partition function is divided into two terms

$$(ap_i)^q + (bp_i)^q, \tag{C.42}$$

so that

$$Z_{k+1} = (a^q + b^q)Z_k. \tag{C.43}$$

Given that $\delta(k) = 3^{-k}$, we obtain

$$D_q = \frac{1}{1-q} \lim_{k \to \infty} \frac{k \log(a^q + b^q)}{-k \log 3} = \frac{\log_3(a^q + b^q)}{1-q} \tag{C.44}$$

In a special case $a = b = \frac{1}{2}$ all D_q are the same and equal to the Hausdorff dimension of a set

$$D_q = \frac{\log_3(2^{-q} + 2^{-q})}{1-q} = \log_3 2. \tag{C.45}$$

For another special case when $a = \frac{3}{4}$, $b = \frac{1}{4}$, the Cantor set is heterogeneous and forms a multifractal with $D_0 = \log_3 2 \approx 0.631$, $D_1 \approx 0.512$, $D_2 \approx 0.428$, $D_\infty \approx 0.262$, $D_{-\infty} \approx 1.262$ [27].

External and internal dimensions of the curve. If we apply the Hausdorff ideas to determine the dimension of the fractal curve γ, and for this reason construct around its points circles of radius $\delta \to 0$, and calculate the area of their association $S(\delta)$ (considering the area of overlapping of several circles only once), then the rate of decrease of the area with δ decreasing will determine the size of the curve [28].

Indeed, for a smooth curve $S(\delta) \sim \delta L$, where L—its length, for a flat area $S(\delta) \sim \delta^0$, and for a fractal its area estimation, which is determined by the width of the neighborhood of a curve, depends in this case not on a radius of a circle, but on a size of curve bends.

Determine the length of the "wave vector" as $k = 1/\delta$, then separate "periods" of curve bends fit in the circle with the size δ, and if $a(k)$ when $\delta \to 0$ decreases more slowly than δ, then circles bandwidth is not of the order δ, but of the order $a(1/\delta)$; when $a(1/\delta) < \delta$, the band of circles manages to track all curve bends, and it is not a fractal, but a smooth [28].

Let the curve γ be fractal. Then, the total area of associations of all circles $S(\delta) \sim a(1/\delta) \sim \delta^\alpha$, and if $0 < \alpha < 1$, then $S(\delta)$ decreases more slowly than for a smooth curve. Consequently, the curve γ occupies an intermediate position between the line and the area. F. Hausdorff proposed a definition according to which the dimension of such a formation is

$$\dim_{\text{ext}} \gamma = 2 - \alpha. \tag{C.46}$$

Index ext (short for external) indicates that when constructing this quantity, we had to go outside the curve itself [28].

By analogy with the two-dimensional case, the external dimension of the fractal curve in space is equal to

$$\dim_{\text{ext}} \gamma = 3 - 2\alpha, \tag{C.47}$$

in n-dimensional—

$$\dim_{\text{ext}} \gamma = n - (n-1)\alpha. \tag{C.48}$$

Let us determine the internal dimension of the curve γ. To do this, we divide it into sections of length δ and introduce a parameter t. We calculate the length of the curve, considering only those of its bends at which t changes by not less than δ. We find that the sum of lengths of these segments is of order $a(1/\delta)\,(1/\delta) \sim \delta^{\alpha-1}$ and tends to infinity with decreasing δ [28].

What is the reason for this divergence. Suppose, following the authors of [28], that we have made a mistake in determining the dimension of our object, and we examine not the curve, but try to define a plane by one parameter instead.

Such parametrization is, of course, bad—the line more and more densely and with self-intersections fills the plane, forming on it the resemblance of a grid. The distance between its bands is $\sim \delta$, and the number of squares is $\sim 1/\delta^2$. Dimensions of polyline segments, densely covering the plane, are small, but their number is very large: the sum of lengths of polyline segments $\sim \delta \cdot 1/\delta^2 \to \infty$. And this is natural: Since the region is two-dimensional, it is necessary to count not its length, but the area, i.e., summarize not lengths of sides, but squares of lengths whose sum is finite [28].

In the case of fractional dimension, it is necessary to summarize some μth degrees of lengths. For the finiteness of resulting sums, we must put $\mu = 1/\alpha$. The dimension of this sum is $\text{cm}^{1/\alpha}$, and the number $1/\alpha$ is itself a dimension. Then, as the internal dimension of a fractal, it is natural to take the number $\dim_{\text{int}} \gamma = 1/\alpha$.

This formula is also true for curves in the space of any number of dimensions.

The external dimension of a curve of the fractal type on the plane varies from 1 to 2 (dimension of the space), the internal one—from 1 to infinity, and they coincide only for the trivial case of a smooth curve. In general, the external dimension of a fractal curve changes from the dimension of a smooth object to the dimension of the space, and the internal one—from the dimension of a smooth object to infinity [28].

"Obviously, in different physical tasks it is necessary to use different definitions of fractal dimension. For example, if we are interested in the task of adsorption on a thin thread, it is important for us to know how many atoms can fit in the neighborhood of the thread, i.e., external dimension. If we want to estimate the weight of the thread, then the internal dimension is important" [28].

Mass dimensions. In the structure of matter, it is always possible to single out the scale which is equal to the correlation length ξ, i.e., to the distance outside of which particles of matter behave themselves statistically independent. What's more, this distance determines the upper boundary of the intermediate asymptotics, and hence the boundary between the intensive and extensive behavior of the density of body's matter. Indeed, when $l < \xi$ the mass of a body is determined by the relation $M \sim l^D$, where D—fractal dimension and by the expression $M \sim l^d$ when $l > \xi$ (herein d—space dimension). Then, the density can be defined as

$$\rho = \begin{cases} M/l^d \sim l^{D-d}, \text{when } l \leq \xi \\ \text{const, when } l > \xi. \end{cases} \tag{C.49}$$

Use this relation and calculate, for an instance, the mass of a sugar cube with an edge L. There is an alternative: To determine the mass through the average density ρ as

$$m = \rho L^3, \tag{C.50}$$

or to take into account that grains of sugar form a statistically self-similar structure, and then the mass is

$$m = \rho_c L^{D_m}, \tag{C.51}$$

where ρ_c—the density of sugar, D_m—mass dimension.

Unlike the "solid" body, in the fractal object the average density depends on the volume, i.e., it is an extensive physical value that decreases as the volume increases. The latter circumstance is easily explained by the rarely used, "non-scientific" definition of a fractal proposed by B. Mandelbrot in a private conversation: A fractal as a physical body is an object in which there are holes of all sizes. Indeed, the average density of a head of Swiss cheese is less than the average density of a piece cut from it: There are fewer holes, and their dimensions are smaller.

From equality

$$\rho L^3 = \rho_c L^{D_m} \tag{C.52}$$

it follows that the exponent D_m, which is equal to

$$D_m = 3 - \ln(\rho_c/\rho)/\ln L, \tag{C.53}$$

determines mass fractal dimension through the true and average density of sugar (which is especially convenient in experimental studies) and, what's more, characteristic dimensions of the body.

Mass dimension D_M could be also defined in a different way. Let it be necessary to calculate the mass of a fractal ball. In this case, the dependence of the mass on the radius behaves like

$$M(R) \sim R^{D_M}, \tag{C.54}$$

where D_M—mass fractal dimension. It is obvious that in the limit

$$D_M = \lim_{R \to \infty} (\ln M(R)/\ln R), \tag{C.55}$$

So certain mass dimensions are global characteristics.

For strictly self-similar mathematical fractals, for example, Sierpinski carpet or Cantor dust, mass dimension D_M coincides with Hausdorff–Besicovitch dimension because it is determined by the dimensionality of the similarity of the scaling law, which is given by the algorithm that generates the fractal [29].

Chemical dimension. If we consider a percolation cluster as a lattice model of a branching polymer molecule, then lattice points correspond to monomers, distances between points—to chemical bonds, and the number of steps in the cluster—the number of chemical bonds along the path of the molecule from the ith monomer to the jth—can be defined as the chemical distance. The dimension associated with this distance is called chemical or the dimension of the coherence D_{ch} [6].

To define D_{ch} let's consider the sphere B_{ch} as a set of points for which $R_{ch} \leq n$, and define chemical dimension as such an indicator that the number of points N, which belongs to B_{ch}, grows as $N \sim n^{D_{ch}}$.

The value D_{ch} is a ratio of two dimensions—fractal dimension of the cluster D and the dimension of the curve D_R, the length of which determines the chemical distance. For the two-dimensional case, the numerically determined value D_{ch} is equal to 1.72, consequently, the dimension of the "geodesic" is $D_R = D/D_{ch} \approx 1.10$; it is clear that it is "not too broken line" [6].

Effective dimension. Effective dimension is the concept that expresses the correspondence between mathematical sets and model objects, and to which, according to B. Mandelbrot, should not be given an exact definition [20].

As a component of the model description, effective dimension has a "special look". It is known that macroscopic objects, even such "puny" as bee's wings, strawberry seeds, a silk thread and a spider web are three-dimensional bodies. But in mathematical models, it can be assumed that wings are of dimension two, the dimension of seeds is zero, the thread—one, and the spider web dimension is between 1 and 2.

The subjective component of effective dimension is clearly visible in the example of Mandelbrot [20]. "Let there be a sphere 10 cm in diameter, twisted from a thick thread with a diameter of 1 mm. To the remote observer, the tangle will appear as a figure with zero dimension, i.e., a point. From a distance of 10 cm, the sphere of threads looks like a three-dimensional body, and from a distance of 10 mm—as a disorderly interlacement of one-dimensional threads. At a distance of 0.1 mm, each thread will turn into a thick rope, and the entire structure will again become a three-dimensional body. At a distance of 0.01 mm, the 'ropes' will turn into an interlacement of fibers, and the sphere will again become one-dimensional, and so on. Finally, when the tangle turns into a cluster consisting of some finite number of points having dimensions comparable to atomic ones, its dimension will again become zero" [20].

Instead of a conclusion. Physical objects and processes can always be associated with a set of dimensions that characterize their properties; the choice of dimensions, naturally, is determined by the purpose of the study and by the essence of the problem being solved. Such an approach, as is known, stimulated research and clarified the essential features of a number of physical systems, allowed to formulate new concepts and models and to achieve a more complete understanding of many long-studied phenomena [6, 7, 20, 28].

References

1. Poincaré, H.: Pourquoi l'espace à trois dimensions. Revue de métaphysique et de morale **20**, 483–504 (1912)
2. Lebesgue, H.: Sur la non applibilité de deux domaines appartenant à des espaces à n et $n + p$ dimensions (extrait d'une lettre à M.O. Blumental). Math. Ann. **70**, 166–168 (1911)
3. Brouwer, L.E.J.: Über den natürlichen Dimensionsbegriff. Journal Für Die Reine Und Angewandte Mathematik **142**, 146–152 (1913)
4. Matematicheskiy entsiklopedicheskiy slovar': Mathematical Encyclopedic Dictionary. Sovetskaya entsiklopediya, Moscow (1988)
5. Hurewicz, W., Wallman, H.: Dimension theory. Princeton University Press, Princeton (1941)
6. Efros, A.L.: Physics and Geometry of Disorder. Percolation Theory. MIR Publishers, Moscow (1986)
7. Sokolov, I.M.: Dimensions and other critical indictors in the percolation theory. Sov. Phys. Usp. **29**, 924–945 (1986), https://doi.org/10.1070/PU1986v029n10ABEH003526
8. Men'shikov, M.V., Molchanov, S.A., Sidorenko, A.F.: Teoriya perkolyatsii i nekotoryye prilozheniya, Rezul'taty nauki i tekhniki. Seriya veroyatnostey. Matematicheskaya statistika. (Percolation theory and some applications, Results of science and technology. Probability series: Mathematical statistics). **24**, 53–110 (1986)
9. Kozlov, S.M.: Geometric aspects of averaging. Russ. Math. Surv. **44**, 91–132 (1989)
10. Men'shikov, M.V.: Otsenki perkolyatsionnykh porogov dlya reshetok v R^n. (Estimates of percolation thresholds for lattices in R^n). Dokl. Akad. Nauk SSSR. (Proc. Russ. Acad. Sci.) **284**(1), 36–39 (1985)
11. Zhikov, V.V.: Asymptotic problems connected with the heat equation in perforated domains. Math. USSR-Sb. **71**, 125–147 (1992)
12. Fizicheskiy entsiklopedicheskiy slovar'. (Physical Encyclopedic Dictionary). Sovetskaya entsiklopediya, Moscow (1984)
13. Schouten, J.A.: Tensor Analysis for Physicists. Clarendon Press, Oxford (1951)

14. Grinchenko, V.T., Matsipura, V.T., Snarskiy, A.A.: Vvedeniye v nelineynuyu dinamiku. Khaos i fraktaly. (Introduction to Nonlinear Dynamics. Chaos and Fractals). LKI, Moscow (2010)

15. Poincare, H.: Science and Hypothesis. Walter Scott Publishing, London (1905)

16. Ehrenfest, P.: In that way does it becomes manifest in the fundamental laws of physics that space has three dimensions? Proc. R. Neth. Acad. Arts Sci. **20**(1), 200–209 (1918)

17. Puga, V.A.: Multidimensional gravitational interaction. The rotation curves of galaxies. JETP **119**(3), 438–450 (2014)

18. Turyshev, S.G.: Experimental tests of general relativity: recent progress and future directions. Phys. Usp. **52**, 1–27 (2009). https://doi.org/10.3367/UFNe.0179.200901a.0003

19. Feder, J.: Fractals. Plenum Press, New York (1988)

20. Mandelbrot, B.: The Fractal Geometry of Nature. W.H. Freeman and Co., San Francisco (1982)

21. Kolmogorov, A.N., Tikhomirov, V.M.: E-entropiya i e-yomkost' mnozhestv v funktsional'nykh prostranstvakh. (E-entropy and ε-capacity of sets in function spaces). Usp. Math. Sci. (Russ. Math. Surv.) **14**(2), 3–86 (1959)

22. Gelbaum, B.R., Olmsted, J.M.H.: Counterexamples in Analysis. Holden-Day, San Francisco (1964)

23. Kahane, J.P., Salem, R.: Ensembles parfaits et series trigonometriques. Hermann, Paris (1963)

24. Pontrjagin, L., Schnirelman, L.: Sur une propriété métrique de la dimension. Ann. Math. **33**, 156–162 (1932)

25. Bozhokin, S.V., Parshin, D.A.: Fraktaly i mul'tifraktaly. (Fractals and Multifractals). R& H Dynamics, Moscow-Izhevsk (2001)

26. Rényi, A.: Probability Theory. North-Holland Publ. Co., Amsterdam (1970)

27. Kuznetsov, S.P.: Dinamicheskiy khaos. (Dynamic Chaos). Fizmatlit, Moscow (2006)

28. Zel'dovich, Ya.B., Sokolov, D.D.: Fractals, similarity, intermediate asymptotics. Sov. Phys. Usp. **28**, 608–616 (1985). https://doi.org/10.1070/PU1985v028n07ABEH003873

29. Schroeder, M.: Fractals, Chaos, Power Laws. Minutes from an Infinite Paradise. W.H. Freeman, New York (1991)

Appendix D
The Relative Order of the Heterogeneous Materials' Structure

Introduction. Numerous studies devoted to the specification of the dependence of the structure and materials' properties, as a rule, contain photographs of surfaces or sections of samples. Such photographs or sections visualize the difference in the structure of composites with different properties. Realizing the importance of the dependence of materials' properties on the structure and understanding that, in fact, properties are the structure expression on their own, the authors of many articles give a verbal description of the photographs, using terms intuitively clear—structuredness, heterogeneity, complexity, but at the same time they do not have clear definitions, calculation algorithms, quantitative component.

In this regard, it is appropriate to have a characteristic that would allow us to quantify the level of structure ordering from photographs. One of the quantities giving this possibility is the relative degree of ordering (RDO) in the substance structure, which is determined from the images [1]—a quantitative characteristic, which is calculated from an operable algorithm based on the representation of the entropy of the information system.

Relative degree of ordering in images. Consider the question of comparing two images as information objects, and determine the relative degree of their ordering [1].

The meaning content and visual perception of the image, as is known, substantially depends among other things on the characteristics of an observing person, on his psychological pattern, mode of thought, etc. [2, 3]. Image formalization is a prerequisite for creating objective characteristics and developing methods of their definition and measurement [4–6]. What's more, when determining the relative characteristic, the question of the correctness of the comparison should be arisen. In case of images, adequate comparison is possible by analogy with the Gibbs theorem, which compares the entropies of equilibrium and arbitrary states [7–9].

© The Author(s), under exclusive licence to Springer International Publishing, part of Springer Nature 2018
A. Herega, *The Selected Models of the Mesostructure of Composites*, SpringerBriefs in Physics, https://doi.org/10.1007/978-3-319-89704-2

The Gibbs entropy is determined by the average value of the logarithm of the equilibrium distribution function of the complete set of coordinates and impulses of system's particles. A comparison of the entropy values of equilibrium and nonequilibrium states can be done only for the same values of the mean energies thereby limiting the set of nonequilibrium states. The conditions for normalizing both equilibrium and nonequilibrium distribution functions must also be the same [7–9].

To determine the RDO of computer images, each of them is being converted programmatically into an array of pixels that is performed in 256 shades of gray. In this case, to ensure the correctness of the comparison, in images the number of pixels and the average amount of the gray level (GL) per pixel must be the same [1].

To satisfy this condition, functions of the density distribution of the gray level values $f_1(i)$ and $f_2(i)$ are recovered from data arrays in pixels of images. And, according to the formula,

$$\tilde{f}_2(i) = f_2(i) \left[\sum_{i=0}^{255} f_1(i) \middle/ \sum_{i=0}^{255} f_2(i) \right] \qquad (D.1)$$

one of them is being renormalized.

Fig. D.1 illustrates the process of GL equalization on two photos by shifting the distribution function of one of them. The abscissa and the ordinate axis on the histograms plot the GL and the number of pixels, respectively. The first photo was brightened, and its corresponding curve in histogram (b) is shifted to the right toward lower values of the GL.

As an illustration, it is appropriate to give the RDO calculated from the GL values for the photos in the Fig. D.1. It is $\Delta S = 5.3588 - 5.0961 = 0.2627$; i.e. the second photo (image) is more ordered in this sense.

Possible options for assessing the relative degree of ordering are linked to the nature of the images formalization: instead of the values of gray color shades there could be constructed others. For an instance, by the ratio, by the difference of values, by the absolute value of this difference

To do this, when forming an array, the boustrophedon as a rule for image bypassing is used (from Greek "The bull way"—the type of writing with variable direction of lines: odd lines are written from right to left, even lines are written in reverse) [10].

Taking into account the capabilities of the program for varying the modes of arrays processing (see Fig. D.2), for each pair of compared images 32 variants of calculating the RDO can be obtained. According to our data, the most sensitive, as a rule, are the estimations made by the GL value in pixels.

Let us show that the Lyapunov function for elements of two equiareal images with the same GL mean value is a measure of the relative degree of ordering (so called I-theorem).

Fig. D.1 Photos and GL distribution histograms before and after renormalzation (on abscissa axis—the gray level, on ordinate—the number of pixels)

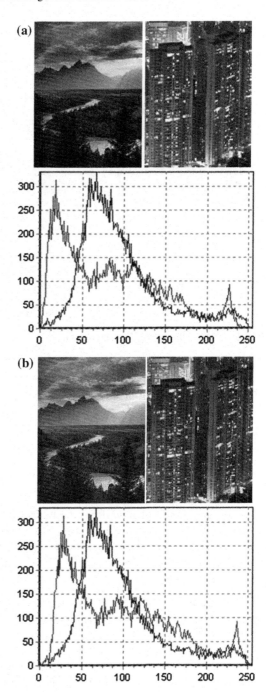

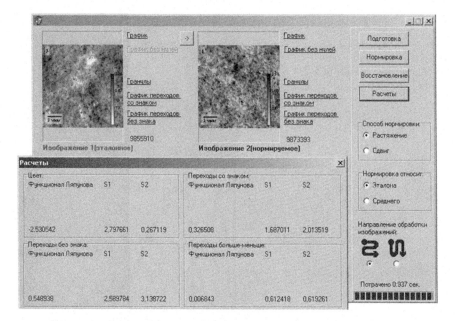

Fig. D.2 The main interface of the program

Let S_1 be the entropy of one image and $\check{S}_2$ be the entropy of the second modified image. Consider the entropy difference. Having left the first two terms in the series expansion of $\ln x = 1 - 1/x$, we have

$$S_1 - \check{S}_2 = -\sum_{i=0}^{255}\left[f_1(i)\ln f_1(i) - \check{f}_2(i)\ln\check{f}_2(i)\right] = -\sum_{i=0}^{255}\left[f_1(i)\ln\left(f_1(i)/\check{f}_2(i)\right)\right].$$

(D.2)

The same result is known from the information theory as the Kullback-Leibler distance [11, 12]: it is a measure of how much two density distributions are far from one another. Furthermore, it is also known in the theory of open systems (S-theorem of Klimontovich—a criterion for the RDO of nonequilibrium states of the open systems [13, 14]).

Two examples. In [15] a study of the structure and properties of oxide coatings deposited on the surface of a deformable aluminum alloy B95T1, which is commonly used for the manufacture of compression-acting structures, by interference microscopy techniques was carried out. The values of some samples parameters were experimentally obtained. It was established, in particular, that the coatings obtained by different technologies have micro-hardness and elastic modulus differing by 15 and 8%, respectively (Table D.1).

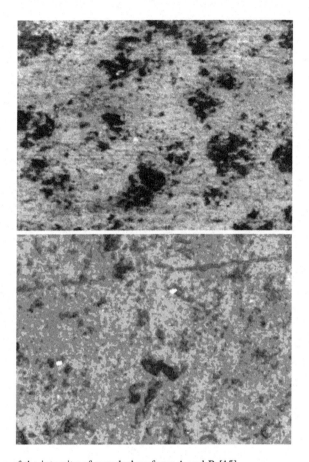

Fig. D.3 Maps of the intensity of samples' surfaces A and B [15]

Table D.1 Mechanical characteristics of coatings (fragment of the table from [15])

Parameter	Sample A	Sample B
Micro-hardness, GPa	10.65 ± 1.6	9.26 ± 2.12
Elastic modulus, GPa	157.2 ± 12.16	170.1 ± 12.4

According to our data, the RDO calculated on the sample surface intensity maps (Fig. D.3) shows that sample A is more ordered in the sense described above—the RDO is 0.26. This means that with an increase in the order of the coating structure its micro-hardness increases, and the elastic modulus decreases (see Table D.1).

In [16] the dependence of the mechanical properties of rubbers containing technical carbon on the "uniformity of the filler distribution in the material" is investigated. The author of [16] believes that "in samples 1 and 3 the distribution can be considered as uniform, and in the second cluster there are clusters in separate groups (see Fig. D.4)". He also believes that "the more heterogeneous the distribution and

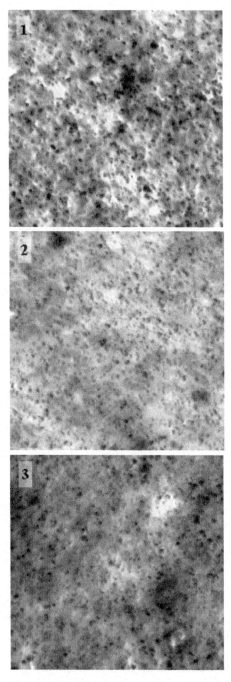

Fig. D.4 Photographs of the filled rubber surface obtained with the help of an atomic force microscope [16]

the larger sizes of clusters, the stronger the filler-filler interaction, the more hetero-geneous stress field and more hysteresis losses during cyclic loading. From this point of view, the second sample has the worst mechanical characteristics" [16].

Using the notion of the RDO of images in the GL distribution showed that the second sample is more orderly than the first ($\Delta S = 4.683$) and third ($\Delta S = 4.752$). Taking into account the results of [16], it is seen that with an increase in the order of the carbon clusters structure in rubber, its mechanical properties deteriorate.

Conclusion

The RDO can be regarded as an integral quantitative characteristic of the physical bodies' structure, which is an alternative to descriptive concepts such as "struc-turedness", "complexity", "heterogeneity".

The possibility of introducing "absolute" order is also evident. To determine it, it is necessary to choose zero of the scale. It can be the image ordering created by a random number generator with a uniform distribution, or, on the contrary, the complete order of the "black square".

References

1. Herega, A.N.: On one criterion of the relative degree of ordering in images. Tech. Phys. **55**, 741–742 (2010)
2. Haken, H.: Principles of Brain Functioning. Springer, Berlin (1996)
3. Solso, R.L.: Cognitive Psychology. Harcourt-Brace, New York (1979)
4. Pratt, W.: Digital Image Processing, vol. 1. Wiley, New York (1991)
5. Soifer, V.A. (ed.): Computer Image Processing. VDM Verlag Dr. Müller, Berlin (2010)
6. Gonzales, R.C., Wood, R.E.: Digital Image Processing. Prentice-Hall (2002)
7. Gibbs, J.W.: Thermodynamics: Statistical Mechanics. Yale Univ, New Haven (1948)
8. Klimontovich, Yu.L.: Statistical Physics. Harwood Academic, New York (1986)
9. Herega A., Sukhanov V., Vyrovoy V.: The model of the long-range effect in solids: evolution of structure clusters of interior boundaries, and their statistical descriptors. AIP Conf. Proc. **1909**, 020069 (2017)
10. Slovar inostrannyih slov. (Dictionary of foreign words). Russkiy yazyik, Moscow (1987)
11. Kullback, S., Leibler, R.A.: On information and sufficiency. Ann. Math. Stat. **22**, 79–86 (1951)
12. Kullback, S.: Information Theory. Wiley, New York (1959)
13. Klimontovich, Yu.L.: Decrease in entropy in the process of self-organization. S-theorem. Sov. Tech. Phys. Lett. **9**, 606–610 (1983)
14. Klimontovich, Yu.L.: Criteria for the relative degree of ordering of open systems. Phys. Usp. **26**, 366–378 (1996)
15. Shestov, S.S., Zagrebin, L.V., Yanovskiy, Yu.G.: Uprochnyayushchiye oksidnyye pokrytiya alyuminiyevykh splavov. Issledovaniya struktury, sherokhovatosti i mikrotvordosti. (Strengthening oxide coatings of aluminum alloys. Investigation of structure, roughness and microhardness). Mehanika kompozitsionnyih materialov i konstruktsiy (Mech. Compos. Mater. Struct.) **16**, 670–679 (2010)
16. Morozov, I.A.: Analiz mikrostruktury napolnennoy reziny pri atomno-silovoy mikroskopii. (Analysis of the microstructure of the filled rubber in atomic force microscopy). Mehanika kompozitsionnyih materialov i konstruktsiy (Mech. Compos. Mater. Struct.) **15**, 83–93 (2009)